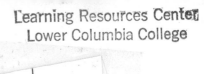

Foundations of Modern Sociology Series

Alex Inkeles, *Editor*

Foundations of Modern Sociology Series

the scientist's role in society

A COMPARATIVE STUDY

Joseph Ben-David, *Hebrew University, Jerusalem*

Prentice-Hall, Inc., *Englewood Cliffs, New Jersey*

Prentice-Hall Foundations of Modern Sociology Series
Alex Inkeles, *Editor*

Current printing (last digit):

10 9 8 7 6 5 4 3 2

PRENTICE-HALL INTERNATIONAL, INC., *London*
PRENTICE-HALL OF AUSTRALIA, PTY., LTD., *Sydney*
PRENTICE-HALL OF CANADA, LTD., *Toronto*
PRENTICE-HALL OF INDIA PRIVATE LIMITED, *New Delhi*
PRENTICE-HALL OF JAPAN, INC., *Tokyo*

C-13-796557-5   P-13-796540-0

preface

This book is an attempt to describe the emergence and the development of the social role of the scientist, and of the organization of scientific work. The subject matter is the social conditions, and, to some extent, the effects of scientific activity, and not the sociology of the contents of scientific knowledge. The topics dealt with are as follows:

1. the social role of the people who created and transmitted scientific knowledge in the traditional societies of antiquity and the middle ages.
2. the problem of why the study of empirical natural science had never, during these periods, become a distinct intellectual role, equal in influence and dignity to the roles of moralists, metaphysicians, lawyers, and specialists in religious learning.
3. the conditions which led to the separation of the role of the scientist from other intellectual roles in seventeenth century Europe, and to the professionalization of science in the nineteenth and twentieth centuries.
4. the principal stages in the development of the organizations of scientific work, such as scientific academies of the seventeenth and eighteenth centuries, the scientific university of the nineteenth and twentieth cen-

turies, and the research institutes and industrial research laboratories starting in the last decades of the nineteenth century.

The subject is treated from a historical and comparative perspective. The emphasis is on the explanation of changes and on turning points, rather than on the interpretation of the ways in which scientific work is conducted in situations of negative stability. Therefore, such topics as the informal structure of the scientific community, the mechanisms of scientific communication and evaluation, and the sociology of the university, of the scientific laboratory, and of scientific teams are not treated under separate headings, but only in the historical context, as conditions or results of change. The framework to which the vicissitudes of the scientific role and of the scientific organization are related is institutional and macro-sociological: namely, the social differentiation, and the educational, political, religious, and economic systems of different societies. While the main purpose of the book is to explore the effects of these conditions on the development of scientific activity, there is also an attempt to conceptualize the effects of the emergence and diffusion of scientific activity in modern societies.

I am indebted to Bernard Barber, R. Bellah, Terry N. Clark, Y. Elkana, S. N. Eisenstadt, and N. Smelser for many helpful comments and suggestions. I have greatly benefitted from collaboration with L. Aran, R. Collins, S. Franklin, and A. Zlochower, and from the assistance of D. Franklin and K. Feinstein in carrying out research on several of the issues dealt with in this book. The Hebrew University of Jerusalem, Israel, The Center for Social Organizational Studies of The University of Chicago, and The Institute of International Studies of The University of California, Berkeley, provided support and facilities for the work on which this book is based. Earlier versions of chapters 4, 6, and parts of chapter 8 were published in *Minerva*. I would like to express my deepest gratitude to the editor of *Minerva*, Edward Shils, for his incisive criticism and invaluable suggestions. Last but not least, I should like to thank Alex Inkeles, the editor of this series, for his helpfulness and forbearance.

Joseph Ben-David

preface

contents

the sociology of
science

one

Approaches to the Study of the Sociology of Science

Sociologists study structures and processes of social behavior. Science, however, is not behavior but knowledge that can be written down, forgotten, and learned again, with its form or content remaining unchanged. Moreover, scientists are engaged in discovering "laws of nature" that cannot be changed by human action. Thus not only are they faced—as in mathematics—with the immanent logic of their own systems of thought, but they accept the further restriction that their systems have to fit the structure of natural events. In principle, this is true also of social scientists and scientific students of culture. They study human behavior and the creation of man as objective things that have discoverable regularities.

Since the subject matter of science is nature and the tools of science are systems of thought, its development is usually conceived of as a history of ideas. This is seen as a series of attempts to explain the working of nature through logically coherent models. The succession of ideas is explained as a result of the discovery of logical flaws within the models or bad fits between the models and the natural events they were supposed to explain.

From such a purely conceptual point of view,[1] there would be little in-

[1] For a concise statement of the case for a history of scientific ideas and against a sociology of science, see A. Rupert Hall, "Merton Revisited, or Science and Society in the Seventeenth Century," *History of Science* (1963), vol. II, pp. 1–15.

terest in the social aspects of scientific work. But there are significant aspects of the development of science that can be systematically explained only by social variables. The value attached to science by society, the interest in making new discoveries contrasted to the preservation of old traditions, the transmission and diffusion of scientific knowledge, the organization of research, and the uses made of science or scientific activity in general are all eminently sociological phenomena.

These considerations suggest a basic criterion for the classification of the literature on the sociology of science, namely, whether it claims that social conditions influence only the behavior of scientists and scientific activity or that they also influence the basic concepts and the logical structure of science. As a second criterion for classifying the literature, I suggest using the kind of variables to which science is related by different authors—whether the variables are predominantly interactional or predominantly institutional. Authors using the interactional approach observe the way scientists act toward each other, such as their division and coordination of work in laboratories, patterns of scientific quotations, and habits of counsultation. The institutional approach relates science to variables that, from the point of view of individual scientists, are given; examples of these variables are the definition of the scientists' roles in different countries, the size and structure of scientific organizations, and different aspects of the economy, political system, religion, and ideology. There is, of course, a great deal of overlap between the two approaches. Whereas the question whether social conditions influence only the behavior of scientists or also shape their concepts is of basic theoretical importance, the choice between the interactional and institutional approaches depends on the problem which is investigated.

Thus, according to our discussion, there are four approaches to the sociology of science: an interactional study of either scientific activity or the conceptual and logical structure of science, and an institutional study of the same two aspects.[2]

The Interactional Approach

With the exception of a recent exploratory work,[3] which attempts to define science exhaustively as the consensus arising from groups of investigators, there have been no attempts at the interactional study of the conceptual and theoretical contents of scientific knowledge. Of the three remaining approaches, the most systematic and concentrated research effort in

[2] This division can be further subdivided. For a systematic overview of the possibilities dealing with the sociology of knowledge in general, see Robert K. Merton, "The Sociology of Knowledge," *Social Theory and Social Structure,* rev. ed. (New York: The Free Press of Glencoe, 1957), pp. 456–488.

[3] John Ziman, *Public Knowledge: The Social Dimension of Science* (New York: Cambridge University Press, 1968), pp. 1–12, especially "(consensus) is the basic principle upon which Science is founded. It is not a subsidiary consequence of the 'Scientific Method.' It is the scientific method itself," p. 9, and the stricture against distinguishing between "science as a body of knowledge, science as what scientists do and science as a social institution," p. 11.

2

the sociology of science today deals with the interactional study of the scientific community or, more concretely, with the network of communication and social relationships between scientists working in given fields or in all the fields.[4] This approach was first employed in the study of scientific productivity of research groups in laboratories.[5] The recent shift of attention from laboratory work groups to networks encompassing distinct fields of research was greatly influenced by the emergence of a view of science as the work of a community in the sociological sense.[6]

This view, first formulated by Michael Polanyi in 1942, has been elaborated recently by Thomas Kuhn.[7] In Kuhn's opinion, scientists in a specific field form

[4] Stephen Cole and Jonathan Cole, "Scientific Output and Recognition. A Study in the Operation of the Reward System in Science," *American Sociological Review* (June 1967), 32:377–390; Diana Crane, "Social Structure in a Group of Scientists: A Test of the 'Invisible College' Hypothesis," *American Sociological Review* (June 1969), 34:335–352; Warren H. Hagstrom, *The Scientific Community* (New York: Basic Books, Inc., Publishers, 1965); Herbert Menzel, *Review of Studies in the Flow of Information among Scientists* (New York: Columbia University Bureau of Applied Social Research, 1958), 2 vols. (mimeographed); Robert K. Merton, "Priorities in Scientific Discovery," *American Sociological Review* (December 1954), 22:635–659; Robert K. Merton, "Singletons and Multiples in Scientific Discovery," *Proceedings of the American Philosophical Society* (October 1961), 105:470–486; Robert K. Merton, "The Ambivalence of Scientists," *Bulletin of the Johns Hopkins Hospital* (1963), 112:77–97; Robert K. Merton, "Resistance to the Systematic Study of Multiple Discoveries in Science," *European Journal of Sociology* (1963), 4:237–282; Nicholas C. Mullins, "The Distribution of Social and Cultural Properties in Informal Communications Networks among Biological Scientists," *American Sociological Review* (October 1968), 3:786–797; Harriet Zuckerman, "The Sociology of the Nobel Prizes," *Scientific American* (November 1967), 217:25–33.

[5] Louis B. Barnes, *Organizational Systems and Engineering Groups: A Comparative Study of Two Technical Groups in Industry* (Cambridge: Harvard University School of Business, 1960); Paula Brown, "Bureaucracy in a Government Laboratory," *Social Forces* (1954), 32:259–268; Barney G. Glaser, "Differential Association and the Institutional Motivation of Scientists," *Administrative Science Quarterly* (June 1965), 10:82–97; Barney G. Glaser, *Organizational Scientists: Their Professional Careers* (Indianapolis: The Bobbs-Merrill Company, Inc., 1964); Norman Kaplan, "Professional Scientists in Industry: An Essay Review," *Social Problems* (Summer 1965), 13:88–97; Norman Kaplan, "The Relation of Creativity to Sociological Variables in Research Organization," in C. W. Taylor and F. Barron (eds.), *Scientific Creativity: Its Recognition and Development* (New York: John Wiley & Sons, Inc., 1963); Norman Kaplan, "The Role of the Research Administrator," *Administrative Science Quarterly* (1959), 4:20–42; William Kornhauser, *Scientists in Industry* (Berkeley, Calif.: University of California Press, 1962); Simon Marcson, *The Scientists in American Industry: Some Organizational Determinants in Manpower Utilization* (Princeton: Princeton University Press, 1960); Donald C. Pelz, G. D. Mellinger, and R. C. Davis, *Human Relations in a Research Organization* (Ann Arbor, Mich., The University of Michigan Press, 1953), 2 vols. (mimeographed); Donald C. Pelz and Frank M. Andrews, *Scientists in Organizations* (New York: John Wiley & Sons, Inc., 1966); Herbert A. Shepard, "Basic Research in the Social System of Pure Science," *Philosophy of Science* (January 1956), 23:48–57.

[6] See Michael Polanyi, *The Logic of Liberty* (London: Routledge & Kegan Paul, Ltd., 1951), pp. 53–57. It was used in the 1950s by Edward A. Shils, "Scientific Community: Thoughts after Hamburg," *Bulletin of the Atomic Scientists* (May 1954), X:151–155, and it became a key concept in the sociology of science in the 1960s. See Gerald Holton, "Scientific Research and Scholarship," *Daedalus* (Spring 1962), 91:362–399; and Derek J. de Solla Price, *Little Science, Big Science* (New York: Columbia University Press, 1963).

[7] See Polanyi, *loc. cit.*; and Thomas S. Kuhn, *The Structure of Scientific Revolutions* (Chicago: The University of Chicago Press, 1962).

3

a closed community.[8] They investigate a well-defined range of problems with methods and tools especially adapted for the task. Their definition of the problems and their methods of investigation derive from a professional tradition of theories, techniques, and skills. And these are acquired through prolonged training that involves, as a matter of fact if not of principle, also some indoctrination. The rules of the scientific method, as laid down by logicians of science, do not, according to Kuhn's view, adequately describe what scientists do. Scientists are not busily engaged in the testing and refutation of existing hypotheses so as to establish new and more generally valid ones. Rather, like people engaged in other occupations, they take it for granted that the existing theories and methods are valid and use these for their professional purposes, which usually are not the discovery of new theories but the solution of concrete problems, such as the measurement of a constant, the analysis or synthesis of a compound, or the explanation of the functioning of a part of a living organism. In his search for a solution, the researcher uses as his model or paradigm the existing tradition of research in his field. He takes it for granted that there is a solution to his problem and therefore regards the problem as a "puzzle."

One of the implications of this is that science is insulated from external social influence, because what scientists consider to be problems and the ways they deal with these problems are determined by their own tradition. It determines which questions can be asked and which are to be excluded, and it defines norms of conduct and criteria of evaluation. The younger scientists are socialized into it; the mature scientists uphold it and transmit it to the next generation. By adopting it, one enters into a community that, like all communities, sensitizes its members to one another and desensitizes them to outsiders. Modern physics, for example, has been the same in the U.S.S.R. as elsewhere, in spite of the totalistic intellectual claims of communism. Even the famous conflict on genetics did not involve a real intrusion of nonscientific criteria in the thinking of the scientific community. Rather, the conflict was a forceful suppression of a scientific community by an autocratic regime instigated by charlatans. Thus, although science is conceived as the activity of a human group ("the scientific community" or, rather, "communities" specialized by fields), this group is so effectively insulated from the outside world that the characteristics of the different societies in which scientists live and work can, for many intents and purposes, be disregarded.

Because the norms and goals of scientific communities are defined by the state of science, their sociology is relatively simple. This, of course, does not make them less interesting. These communities can serve as an example of an extreme case of effective social control by a minimum of informal sanctions.

[8] The word *community* does not distinguish between the different kinds of social bonds. An early attempt to make such a distinction and identify the bond characteristic of religious and social groups that are held together somewhat similarly to the scientific community was that of Herman Schmalenbach, "Die soziologische Kategorie des Bundes," *Dioskuren* (1922), 1:35–105. For a recent treatment of the general question, see Edward Shils, "Primordial, Personal, Sacred and Civil Ties," *British Journal of Sociology* (1957), 8:132–134.

the sociology of science

They comprise one of the interesting instances where a group of people is held together by a common purpose and shared norms without the need of reinforcement by familial, ecological, or political ties.

However, this scheme, which Kuhn calls "normal science," does not in his view explain scientific change, which explanation is his main aim. Kuhn conceives of scientific change as a series of "revolutions." Every paradigm sooner or later reaches a point of intellectual exhaustion. Some puzzles persist and resist solution, and after a while the conviction gains ground that they cannot be solved with the aid of the existing models of theory and procedure. There then arises a crisis within the scientific community similar to that which arises in any community when the goals that inspire it become unattainable by the accepted means. This is the condition that sociologists call *anomie* (normlessness); it has been studied widely as the background of social deviance and change.[9]

According to Kuhn, in such periods of crisis the barriers between science and the broad intellectual currents of society break down. In their search for a basically new orientation, scientists in a field in crisis become interested in a variety of philosophical ideas and theories far removed from their own specialty. There is no longer a consensus concerning the correct approach to problems, and it is impossible to predict which thought model derived from where will provide the starting point for the rise of a new paradigm.

The main intent of the concept of scientific revolution is philosophical— to show that the development of science is not cumulative, as it is usually considered to be, but consists of a series of distinct and disconnected beginnings, growths and declines, somewhat like the rise and fall of civilizations. Carried to its extreme, this view would, for instance, deny any continuity between the concepts and standards of solution accepted in classical and current physics— this is a position difficult to accept.[10]

From a sociological point of view, the assertions that revolutions regularly follow the exhaustion of paradigms and occur neither before nor after that point and that, furthermore, revolutions are completely discontinuous and different from other types of change would make the scientific community a social anomaly. This extreme view of revolution is a postulate made necessary by the assumption that *normally* scientists work within existing paradigms. Hence the abandonment of an existing paradigm and the creation of a new one can occur only where the paradigm actually breaks down. Empirically, however, there may be (*a*) differences among individuals and groups in their perceptions of the

[9] Emile Durkheim, *Suicide* (New York: The Free Press of Glencoe, 1952), pp. 241– 276; Robert K. Merton, "Social Structure and Anomie," *Social Theory and Social Structure*, 2nd edition, pp. 131–194; Talcott Parsons, *The Social System* (New York: The Free Press of Glencoe, 1951), pp. 256–267, 321–325.

[10] Dudley Shapere, "The Structure of Scientific Revolutions," *Philosophical Review* (July 1964), LXXIII:383–394. A modified version of Kuhn's theory which appeared after the completion of this manuscript eliminates this difficulty, as well as those raised in the next paragraph: Thomas S. Kuhn, *The Structure of Scientific Revolutions*, Second Edition (Chicago: University of Chicago Press, 1970), pp. 176–207. This modified version is consistent with the sociological interpretations contained in these pages.

breakdown (or exhaustion) of the paradigm due to either their different locations in the scientific community or differences in their individual sensitivity and (b) differences in the closure of certain scientific communities, that is, some may have nothing to do with other scientific communities whereas others may have partially overlapping interests and common personnel. It is possible, therefore, to envisage normative variation leading to as fundamental a change as a revolution but issuing from the feelings of frustration and search for innovation of only a portion of the scientific community.[11] This, of course, also implies that paradigmatic behavior is a limiting state, which scientific communities tend to approach but which they never actually attain.

The ideal typical description of this limiting state has nevertheless been very useful in the conceptualization of the scientific community, which can be identified as a group that tries to behave as if it followed a commonly agreed on and stable paradigm. Although, in fact, there is a great deal of variation and constant change in the contents of science, the assumption of the existence of paradigms helps to define the boundaries of a community, just as assumptions about the existence of other kinds of common traditions define the boundaries of national, religious, and other spatially nondistinct groups.

In any case, whether one explores the statics or the dynamics of the scientific community thus conceived, the investigation is, as pointed out, strictly interactional. The growth of scientific knowledge and changes in scientific interest are related to the activities of a network of scientists working in a field. Coordinated advance on a common front is related to proper exchange of information and rewards, expressed in short time gaps between publication and citation and in practices of recognition and honor that are closely related to objective indexes of merit. Changes in the interests and goals of a whole or, more usually, a part of a scientific community are related to breakdowns in interaction or communications. These breakdowns are due to a variety of circumstances, ranging from a simple overloading of the network when it grows beyond a certain size to basic innovations that arise in response to the inadequacies of the existing tradition for the purposes of part or all the community.

The Institutional Approach

Whereas the interactional study of science has concentrated on the explanation of the behavior and activities of scientists but rarely touched on the contents of scientific knowledge, the institutional tradition has concentrated to a large extent on the latter. One institutional explanation of the contents of scientific knowledge is directly related to the concept of scientific revolutions. Normally the contents of science are defined by the existing scientific tradition, but when more basic scientific change occurs, the tradition is partly overthrown. The closure and the specificity of the disciplines

[11] For some elaborations of these points, see Joseph Ben-David, "Scientific Growth: A Sociological View," *Minerva* (Summer 1964), 3:471–475; and Hagstrom, *op. cit.*, pp. 159–243.

6

are abolished, and the scientific community is open to outside influences. Of course, these may still come from science. In the investigation of certain phenomena, existing traditions of one field may be far-reachingly modified by those of another one; for example, methods developed in physics may be applied to chemical analysis or those of chemistry to the understanding of physiological processes.

It has been asserted, however, that the ideas leading to basic scientific change are often derived from general, nonscientific metaphysical speculation. Thus the transitions from Aristotelian to Newtonian physics and from classical Newtonian to modern physics are not explicable by the immanent logic of scientific thought and empirical verification. The new theories were not implied in those that preceded them. Rather, the preconditions for the development of the new physics in both cases were (a) the abandonment of the existing view of nature accompanied by the rise of an interest in a broad range of basic philosophical questions and (b) the rise of a new view of science (or of an appropriate part of it) using different concepts and methods from those of the old one. On such occasions, basic scientific change involves science with broader intellectual currents.

The most influential source of this view was Alexandre Koyré, who explored the influence of Platonic philosophy on the foundation of classical (Newtonian) physics. In his interpretation, the rise of the latter was part and parcel of the anti-Aristotelian movement in philosophy.[12] Similar explanations were offered for the rise of modern physics, the electromagnetic theory, and thermodynamics. It has been suggested that Faraday and Oersted both came to their ideas about the structure of electromagnetic fields under the influence of the holistic views of the *Naturphilosophie* and that Helmholtz's formulation of the energy concept was influenced by Kant.[13]

All this does not imply external influence on science. Since the seventeenth century, philosophies to a large extent have been attempts to explore the underlying logic of the sciences, to apply scientific principles to moral problems, or to differentiate the fields where scientific logic applies from those where it does not. Because this philosophy is constantly influenced and challenged by science, the influence is, at times, inevitably reversed.[14]

But there is one area of sociology, the so-called sociology of knowledge, that asserts that there are regular relationships between the perspectives and motives of social groups on the one hand and philosophical, legal, and religious

[12] Alexandre Koyré, *From the Closed World to the Infinite Universe* (New York: Harper & Row, Publishers, Incorporated, 1958).

[13] Pierce Williams, *Michael Faraday* (London: Chapman & Hall, Ltd., 1963), pp. 60–89. On thermodynamics, see Yehuda Elkana, *The Emergence of the Energy Concept*, doctoral dissertation, Brandeis University (Ann Arbor, Mich.: University Microfilms, Inc., 1968, no. 68–12, 434).

[14] For a systematic exploration of these mutual influences in the nineteenth century, see Stephen Brush, "Thermodynamics and History," *The Graduate Journal*, vol. 7:2 (Spring, 1967), pp. 477–565.

7

(or ideological) systems of thought on the other. Although natural science, which is not concerned with human affairs and experiences, has not been considered to be directly determined by social perspective and motives, it may be so determined indirectly by the latent unstated philosophical premises of science.[15] According to this trend of thought, social determination of science depends on (*a*) the existence of a systematic relationship between the conceptual structure of philosophies prevailing at the time, on the one hand, and variables of the social situation, on the other and (*b*) a systematic relationship between those philosophies and science. It should be emphasized that both relationships must be systematic, that is, regular and predictable. Occasional influences may provide the theme for historical investigations but not for a sociology of science.

It appears, however, that none of these relationships is systematic. Let us consider one of the best known and apparently most reasonable of the hypotheses concerning the relationship between the contents of a philosophy and social structure, that is, that liberalism as a social philosophy is related to the existence of a powerful merchant class (bourgeoisie). The usual form in which this hypothesis is stated is so general that it is meaningless. Bourgeoisie is defined as including liberalism (*bourgeois liberalism*), so the existence of the relationship becomes a foregone conclusion.[16] But it is possible to extract specific testable relationships from this generalization. One of these stresses that the individualism and rationalism of a liberal philosophy result from the interest of merchants in calculability and in the definition of human relations in terms of economic transactions. Therefore, involvement in a capitalistic economy creates a predisposition to view society in an atomistic manner, as the total of all individuals acting on the basis of considerations of means and ends, rather than as an organic entity steeped in tradition and primordial experience of the group, which precedes the individual and of which the individual is only a part.[17]

If this hypothesis were correct, individualistic philosophies would be in favor of, or at least consistent with, policies designed to promote the interests of merchants, and merchants would have favored those philosophies. But this is not the case. One of the first and most significant individualistic philosophers, Thomas Hobbes, was an advocate of absolutist monarchy. On the other hand, to Adam Smith, the most important economist of the eighteenth century and

[15] This approach was current among some Soviet philosophers and scientists in the 1920s. David Joravsky, *Soviet Marxism and Natural Science, 1917–1932* (London: Routledge & Kegan Paul, Ltd., 1961).

[16] Karl Mannheim, *Ideology and Utopia* (London: Routledge & Kegan Paul, Ltd., 1946), pp. 108–110.

[17] The most influential source of this view has been Karl Marx. See Karl Marx and Friedrich Engels, *The Communist Manifesto* (Harold J. Laski, ed.) (London: George Allen & Unwin, Ltd., 1954). For a detailed exposition of the Marxist view, see George Lukacs, *Geschichte und Klassenbewusstsein* (Berlin: Der Malik Verlag, 1923), pp. 102–103, 144–145, 148–149.

8

the sociology of science

certainly an individualistic philosopher, the merchants were always suspect of seeking for themselves monopolistic privileges. This should be enough to cast serious doubts on the assumption that class bias in any way determined the perspectives of philosophers. There is also no evidence that merchants were systematically in favor of individualistic (or any other) philosophies. They were interested in profit and were willing to support any policies that seemed likely to increase profit. And their support was usually based on very short-range considerations, not on philosophy.

It is similarly difficult to relate the modern collectivistic philosophies to class interests or perspective. These philosophies derive mainly from Rousseau, one of the great intellectual influences on the French bourgeois revolution. Subsequently, collectivism appeared in the conservative thinking of Hegel in Germany and Bonald and De Maistre in France and reappeared one generation later in the "progressive" philosophies of Comte and Marx.

It appears, therefore, that there is no relationship between class interests on the one hand and the concepts and methods of philosophy on the other, even in the so-called ideological field, where such a connection appears highly plausible.[18] There probably is some relationship between the concrete social or cultural problems that philosophers devote their attention to and the surrounding social reality, but if so, this relationship is trivial. Except in the completely deductive fields of mathematics, people theorize about what they observe. And in social thought, what could be observed meant until recently only one's immediate surroundings. Now, as a result of polling techniques, systematic compilation of statistics, and the numerous possibilities of observing social situations that are alien to the background of the observer, even this limitation has been greatly reduced.

This is not to deny that on occasions philosophers are influenced by their social preferences. But these are usually just instances of poor philosophy, often a kind of obiter dicta which have little relationship to the theoretical part of the philosophies of otherwise good philosophers. This can be shown by some examples. Hobbes's atomistic model was an attempt to diagnose social and political disintegration by applying a model borrowed from natural philosophy. His solution may have reflected his own preferences, but a person with quite different preferences, such as John Locke, could apply essentially the same philosophical model for the conceptualization of quite a different political society. The idea was not new, and its systematic application was probably influenced by physical theories current at the time and by its usefulness in analyzing economic and political processes in nonfamilistic and nonreligious societies.

[18] In his historical writings, Marx was aware that people did not actually behave according to his theories. He tried to circumvent the difficulty by stating that the philosophical representatives of a class interest do not necessarily belong to it, that they only express views that reflect the activities of a given class. This takes away all the theoretical significance of his argument about the class determination of philosophy (ideology). Merton, "The Sociology of Knowledge," *op. cit.*, pp. 463–464.

the sociology of science

To give another example, Marx's philosophy can probably be best explained as an attempt by a Hegelian philosopher (never able to unlearn the philosophy he had been taught at school) to assimilate to this philosophy an entirely different kind of intellectual tradition (English economics). Given Marx's intellectual background, this was an important and vital problem for him, and in his search for a solution he made use of whatever observations were available to him. In these observations, as well as in the economic theories with which he grappled, the problem of labor loomed large. Eventually Marx's personal career became closely tied to the socialist movement, the rise of which preceded his philosophy. But whether his philosophy represented the actual interests of the working classes is at least a debatable question. And there is certainly no evidence or even a shred of a reason to believe that anything like Marxist philosophy would have arisen in either England or France, which had had important industrial working classes and socialist movements. Thus there can be no doubt that Hegelian philosophy and English economics were necessary preconditions of Marxian philosophy and that the economic situation of industrial workers was one of the observations with which it dealt. But it is impossible to show that any important class or political interests had anything to do with the genesis of its basic concepts, methods, or theories.

If this argument is correct, the possible relationships between social structure and philosophy are greatly reduced. They may consist of the application of existing theories to problems that are acute in a given society. Or, to the extent that there are competing theories, the relationship may consist of the choice of those that seem to be more relevant to the problems. And finally the relationship may consist of any modifications of theories or theoretical innovations arising from the attempt to understand a situation to which the available concepts are not adequate. Thus even if it could be shown that speculative systems of philosophy have systematically influenced science, in the majority of cases this would not imply the existence of any systematic social influence on science. Social conditions are reflected in the substantive contents of philosophical discussions but only rarely in the concepts and theories of the different philosophies (and then too, the social conditions reflected may be those of a distant past). What can, however, influence science is the conceptual and theoretical structure, not the contents.

Furthermore, even if this conclusion were false and philosophies are reflections of social conditions, there would still be doubts about the systematic influence of philosophy on science. This can be shown by the example of *Naturphilosophie*, which was referred to above. The influence of this philosophy on certain scientific theories might have been real, but neither its choice by some scientists nor its effects on their work can be attributed to any systematic factors. If *Naturphilosophie* proved fruitful in physics, it was because at the time, the internal state of physics happened to be such that some holistic approach was useful in the solution of certain problems. But this does not mean

10

that *Naturphilosphie* furnished the actual concepts or methods.[19] Faraday and Oersted used the philosophical ideas only to the extent that they helped to remove the barriers to advancement presented by a physical theory that was considered nearly closed and perfect. Had they tried to establish some systematic unity between physics and *Naturphilosophie*, they would have failed dismally, as did all those who tried to do so. In fact, the attempt to use this philosophy systematically led to the failure of German biology (the development of which had been seriously hindered for about two decades by the sway of *Naturphilosophie*) and here and there damaged chemists too.[20] Advancement in both these fields came only after scientists abandoned *Naturphilosophie*.

Darwin's evolutionary theory, the most important innovation in mid-nineteenth-century biology and the one in which the influence of social thought has been most unequivocally traced, made important use of the ideas of competition and selection as conceived by the economists, especially Malthus.[21] These ideas were derived from an individualistic model of society, related to the analytical atomistic philosophies of the seventeenth and eighteenth centuries. Thus, in the same period when holistic philosophies seemed to provide inspiration for new thinking in physics, the sciences of biology and chemistry were more fruitfully inspired by atomistic philosophies. This shows that the question of which philosophy was or was not useful to scientific growth depended on (*a*) the state of the particular science and not some common underlying state of social affairs or spiritual culture and (*b*) the discernment of the scientists in using the philosophical ideas in contexts determined by the problems inherent to their scientific specialties.

It can be concluded, therefore, that although ideological bias (socially determined or not) might have played some role in the blind alleys entered by science, the philosophical assumptions that had become part of the living tradition of science were selected by scientists from the array of competing philosophies for their usefulness in the solution of specific scientific problems and not for any socially determined perspective or motive. The scientists borrowed from the philosophies points of view or hunches for looking at a problem from a new angle but did not adopt the philosophical systems.

Finally, there is the possibility that social situations influence the course of science (like that of social thought) by drawing attention to certain subjects rather than others. Certainly political and economic pressures have directed

[19] Kant's philosophy probably did contain useful concepts, but these were directly derived from and related to natural science. They cannot be regarded, therefore, as an external influence on science. See Elkana, *op. cit.*

[20] Richard Harrison Shryock, *The Development of Modern Medicine: an Interpretation of the Social and Scientific Factors Involved* (London: Victor Gollancz, Ltd., 1948), pp. 192–201; Wilhelm Prandtl, *Humphrey Davy—Jöns Jacob Berzelius Zwei Chemiker* (Stuttgart: Wissenschaftliche Verlagsgesellschaft M.B.H., 1948), pp. 117–253.

[21] See, however, Gertrude Himmelfarb, *Darwin and the Darwinian Revolution* (Garden City, N.Y.: Doubleday & Company, Inc., Anchor Books, 1962), pp. 159–167 for qualifications concerning this influence.

the attention of scientists to certain important practical problems, but the effect has been much more limited than is usually believed. Perhaps the best example of this is provided by the grandiose efforts in the U.S.S.R. during the last fifty years to direct science. These efforts led to the training of large numbers of scientists, and as a result, the general level of scientific activity in that country has been raised. But there is no indication that the attempts at the selective development of certain fields created a science which is different from that of other countries. The excellence of Soviet physics could be quoted as evidence of the success of directing attention to a militarily important field. However, Soviet physics has not been different from other physics, so all that its preferment by the government made possible was the relatively efficient exploitation of the potentialities inherent in the general state of the discipline. In fields such as plant genetics, where science could not deliver the desired solutions, or economics, where the solutions were inconsistent with political purposes, the attempt to force scientists to produce results led only to scientific decline and stagnation.[22] Thus although societies can accelerate or decelerate scientific growth by lending or denying support to science or certain parts of it, they can do relatively little to direct its course. This course is determined by the conceptual state of science and by individual creativity—and these follow their own laws, accepting neither command nor bribe.

Another trend in the sociology of knowledge seeks to establish the relationship between the economy and science through the mediation of specific technologies. According to this view, economy sets the tasks for technology, and technology, in turn, poses problems or suggests solutions for science.[23] For example, it seems plausible to seek connections between the revolution in astronomy during the sixteenth and seventeenth centuries and the concern with problems of navigation during that period, just as there is an obvious connection between the hot and cold wars of the last decades and the development of nuclear physics and space exploration. However, although circumstances may have increased the supply of scientists and thus sped development in the related fields, there is no evidence whatsoever that these circumstances significantly influenced the contents of scientific ideas. There is even some doubt about the simple relationship between the demand for certain types of knowledge for practical purposes and the volume of relevant scientific activity in any given country. For instance, the Spanish and Portuguese, who were among the leading seafaring peoples during the period of growth of the new astronomy, had contributed little to the development of that astronomy, whereas the

[22] David Joravsky, "The Lysenko Affair," *Scientific American* (November 1962), CCIX 41–49.

[23] Boris M. Hessen, "The Social and Economic Roots of Newton's Principia," in George Basalla (ed.), *The Rise of Modern Science* (Boston: D. C. Heath and Company, 1968), pp. 31–38; Edgar Zilsel, "The Sociological Roots of Science," *American Journal of Sociology* (January 1942), XLVII:544–562; J. D. Bernal, *The Social Functions of Science* (London: Routledge & Sons, 1939).

the sociology of science

partially landlocked Poles and Germans played a central role, because the ideas of Copernicus and Kepler set the framework for the scientific revolution.

Similarly, the developments in nuclear research were not responses to technological demand. The relative backwardness of nuclear research in Germany during World War II was partly due to reasons similar to those that account for the decline of Spanish-Portuguese astronomy. In both cases science-based innovations failed to develop because the countries did not provide the scientists with the conditions necessary for the maintenance of their autonomy. The fact that there was technological demand for the innovations was not a sufficient condition for their emergence. In any case, all the developments in nuclear physics necessary for the production of the atomic bomb had preceded the plans for its actual production. Even the type of research organization required for this field had been started in the thirties (by Lawrence at Berkeley). It is true that since World War II the research of subatomic particles has been greatly furthered by financing that became available because of the original practical applications of atomic research. But the results of this expensive research have had no practical application, and this shows how tenuous is the relationship between practical purpose and scientific theory.

Unlike the relationship between science and philosophy, the relationship between science and technology is reciprocal—scientific thought is not regularly and predictably influenced by the needs of technology, and technology is not regularly and predictably influenced by science.[24] This is less surprising than it may appear, because after all technological application depends on profitability. The creation of applicable knowledge is not, therefore, a sufficient condition of its technological exploitation. It only creates an opportunity for such use, but it cannot determine the time (except its lower limit) and the place this will occur. Besides, a considerable part of the technological inventions is not based on scientific knowledge but on intuition and practical experience.

Technology creates opportunities for science through the invention and production of instruments. Scientific investigations require tools, as much as does industrial production. And tools devised for the latter may also serve the purposes of the former, or vice versa. Furthermore, some of the tools of science can be produced only by an advanced industry, and the support of big science requires investments that only a large economy can provide. These relationships are self-evident. But none of them implies that scientific ideas are *determined* by economic interests either directly or through the mediation of technology.

Conclusion

We have seen that although there is a possibility for an interactional sociology of scientific activity, the possibilities for either an interactional or institutional sociology of the conceptual and theoretical contents of

[24] Jacob Schmookler, *Invention and Economic Growth* (Cambridge, Mass.: Harvard University Press, 1966).

science are extremely limited. The remaining approach, an institutional sociology of scientific activity, is the one that will be followed in this book.[25] We shall examine the conditions that determined the level of scientific activity and shaped the roles and careers of scientists and the organization of science in different countries at different times.

Science and Economics

Before going further, it should be stated how this institutional sociology is related to the economics of science. After all, such things as the "level of scientific activity" and important aspects of careers and organizations inevitably require economic resources. This seems to indicate that any sociological investigation of these problems would explicitly have to include economic conditions. However, for the overwhelming part of our study, the economic conditions can be regarded as given. There are a number of reasons for this.

In order to treat an activity—in this case, science—in terms of economic exchange, there must be some differentiation between supply and demand. But for the period preceding the seventeenth century, science was a very small-scale activity. It usually consisted of little more than private individuals' occasionally observing the skies to account for the movement of the stars and informally exchanging views on the subject with their friends who had similar interests or conducting similiar amateur activities in other fields. Thus an economic investigation of the subject in this period would be about as useful as an attempt at the economic analysis of private prayer or neighborhood gossip.

Only since the latter part of the seventeenth century has there been a discernible social demand for science. Since then, economic investment in science has probably grown continuously, and, in this general sense, science has become part of the national economy. For this reason, scientific activity has taken place in economically advanced countries and not elsewhere.—A certain

[25] This aspect is treated also in Bernard Barber, *Science and the Social Order* (New York: The Free Press of Glencoe, 1952); Joseph Ben-David, *Fundamental Research and the Universities* (Paris: OECD, 1968); Diana Crane, "Scientists at Major and Minor Universities: A Study of Productivity and Recognition," *American Sociological Review* (1965), 30:699–714; Renée C. Fox, "Medical Scientists in a Château," *Science* (November 5, 1962), 136:476–483; Renée C. Fox, "An American Sociologist in the Land of Belgian Medical Research," in P. E. Hammond (ed.), *Sociologist at Work* (Basic Books, Inc., Publishers, 1964), pp. 345–391; Robert Gilpin, *France in the Age of the Scientific State* (Princeton, N.J.: Princeton University Press, 1968); Norman Kaplan, "The Western European Scientific Establishment in Transition," *The American Behavioral Scientist* (1962), 6:17–21; Robert K. Merton, "Science, Technology and Society in Seventeenth Century England," *Osiris* (1938), IV:360–632; Parsons, *op. cit.*, pp. 335–348; Don K. Price, *Government and Science: Their Dynamic Relation in American Democracy* (New York: New York University Press, 1954); Don K. Price, *The Scientific Estate* (Cambridge, Mass.: Harvard University Press, 1965); Derek J. de Solla Price, *Little Science, Big Science* (New York: Columbia University Press, 1963); Norman W. Storer, *The Social System of Science* (New York: Holt, Rinehart & Winston, Inc., 1966); Alvin M. Weinberg, *Reflections on Big Science* (Cambridge, Mass.: M.I.T. Press, 1967).

level of wealth was a necessary but not a sufficient condition for its emergence. Furthermore, following the emergence of modern science differences in wealth (above a certain level) do not explain the differences in the scientific conrtibutions of the different countries.

Although it is impossible to find a satisfactory way to measure the scientific contributions, it is possible to discern the outlines of the ecology of modern science since its first stirrings in the sixteenth century. From the writings about science and the itineraries of advanced students and researchers, it is evident that from the beginning, scientific activity tended to be disproportionately centered in one area. Until the middle of the seventeenth century, the center for all scientific study was Italy,[26] but in the second half of the century, the center shifted and everyone who was interested in science wrote and spoke about its favorable situation in England.[27] However, because developments in France closely followed those in England, Paris became the undisputed center around 1800. No scientist could afford not to read and speak French and all went to Paris to study, to do research, or just to meet the most famous people in their fields. Forty years later, the meeting and training place for the scientists of the world became Germany; that country retained its position until the 1920s.[28] Then the center shifted to the United States, with Britain holding a secondary position.[29]

There is no satisfactory way to quantify this information, although the amount of time spent by advanced students of science in countries other than their own would probably reflect these shifts fairly precisely. And in this century the distribution of Nobel prize winners also seems to be a good indication of where the center of research has been. Other indexes that reflect these changes, such as numbers of publications, discoveries, and scientists, can be tabulated for earlier periods (see Appendix).

A comparison of the geographical shifts of scientific activity with historical information about the wealth of the various countries does not show any indication that scientific growth was the result of economic growth. It is possible that the shift of the scientific center from Italy to England was connected with changes in the economic importance of these two countries. But there was no relation at any time during the sixteenth and seventeenth centuries between the economic and scientific positions of either Spain or Portugal. Neither the scientific predominance of France early in the nineteenth century nor that of

[26] Harcourt Brown, *Scientific Organizations in Seventeenth Century France (1620–1680)* (Baltimore: The Williams & Wilkins Company, 1934), pp. 3–6.
[27] *Ibid.*, pp. 119–128, 145–147, 216–217. These pages contain a description of French seventeenth-century pamphlets that propagandized for public recognition and support of science. All of the pamphlets refer to England as a model.
[28] H. I. Pledge, *Science Since 1500* (London: H.M. Stationery Office, 1947), pp. 149–151; Donald Stephen Lowell Cardwell, *The Organization of Science in England: A Retrospect* (London: William Heinemann, Ltd., 1957), pp. 50, 106, 134–136.
[29] Charles Weiner, "A New Site for the Seminar: The Refugees and American Physics in the Thirties," in *Perspectives in American History*, Vol. II, 1968, pp. 190–223; Ben-David, *Fundamental Research and the Universities.*

15

Germany in the middle of the century appears to be a result of their economic position, and it was some seventy years before the scientific position of the United States matched its rank among the wealthy countries of the world.[30]

This geographic shifting suggests that there probably is some connection between economic and scientific growth, but that it is not a direct connection. More probably, the two are related by a common underlying characteristic, such as talent, the social motivation of advancement, or something similar. Wealth is, of course, necessary for research. But until the 1950s, the amounts spent on research were such a small part of the economy of any nation that all the wealthy countries could easily afford to compete (see the Appendix).

As can be seen from Table 8 in the Appendix, even today there are some doubts about the relationship between the investment of men and money in science and scientific production, especially high quality production. Of course this is not surprising in a field where the quality and training of the workers play such a decisive role.

An Overview of the Book

These considerations clear the way to the questions discussed in this book. Chapters two to four deal with the conditions that prevented science from becoming a socially valued activity in all the types of human society except a single and rather late one. The conditions that made possible the emergence of science in that society are also examined.[31]

The main sociological concept to be used in these chapters is that of the "role." This is the pattern of behaviors, sentiments, and motives conceived by

[30] W. A. Cole, "The Growth of National Incomes," in H. J. Habakkuk and M. Postan (eds.), *The Cambridge Economic History of Europe* (New York: Cambridge University Press, 1966), Vol. VI, pp. 1–55. The tentative findings of Derek J. de Solla Price, "Measuring the Size of Science," unpublished lecture delivered at the Israel Academy of Science, November 2, 1969, show a fairly good correlation between the share of different countries in the world's economy and their share in the production of scientific papers and productive scientists. But this can still not be considered as evidence of a causal relationship. The findings may indicate the emergence of such a causal relationship between science and the economy, but they may also be the result of the wider and more efficient diffusion of the patterns prevailing at the center. As a result of social conditions, there emerges in the center a certain level and certain forms of scientific activity. And because the center influences scientists all over the world, they use it as a model for the organization of science in their respective countries. But the scientists will be successful in emulating the central model only within the limits determined by the wealth of their countries, because it is difficult to persuade governments to spend relatively much more on science than the model country spends on it. Thus the money spent on science in each country will be a fairly uniform percentage of its gross national product, not because economic activity determines or is determined by science but because of the way all countries imitate one.

[31] The first systematic study of the sociological aspects of this problem was made by Merton, *op. cit.*, 1938. The historical literature on the subject is vast. For a recent summary see Marie Boas, *The Scientific Renaissance: 1450–1650* (New York: William Collins Sons & Co., Ltd., 1962); and A. Rupert Hall, *From Galileo to Newton, 1630–1720* (New York: William Collins Sons & Co., Ltd., 1963).

16

the sociology of science

people as a unit of social interaction with a distinct function of its own and considered as appropriate in given situations. This concept implies that people understand the purpose of the actor in a role and are capable of responding to it and evaluating it. The persistence of a social activity over long periods of time, regardless of changes in the actors, depends on the emergence of roles to carry on the activity and on the understanding and positive evaluation ("legitimation") of these roles by some social group.[32] In the absence of such a publicly recognized role, there is little chance for the transmission and diffusion of the knowledge, skills, and motivation pertaining to a particular activity and for the crystallization of all this into a distinct tradition.

Thus the existence of people interested in the regularity of heavenly phenomena, or in the characteristics of plants and animals, or any other question defined today as scientific could not in itself give rise to a scientific tradition. Where these interests were not considered as integral parts of any role, there emerged hardly any tradition at all. Traditions developed only where such knowledge was considered as part of different roles: astronomy as part of the priestly role; knowledge of plants as something appropriate to farmers; and that of animals as useful for hunters or herdsmen. But there was no tendency to subsume such knowledge under abstract laws or often not even under any laws, since they were not considered intellectual concerns but technical information.

The discussion in chapter four will therefore show how the scattered interests and activities related to the understanding of natural events evolved into the publicly recognized role of the scientist.

The emergence of a new social role takes place within a more comprehensive social setting. According to the definition given here, its very emergence implies a change of social values. In the case of the scientific role, that change in values meant the acceptance of the search for truth through logic and experiment as a worthwhile intellectual pursuit. This modified philosophical and religious authority, raised the dignity of technological knowledge, created new conceptions and norms concerning intellectual freedom in general, and eventually had far-reaching effects on practically all the traditional social arrangements. Therefore, the emergence of the scientific role was connected to changes in the normative patterns ("institutions") regulating cultural activity, and also (subsequently and indirectly) to other kinds of social activity. This institutional change which, like the emergence of the scientific role, first occurred in England, will be discussed in chapter five.

These antecedents set the stage for the development of the organization of science and the scientific community. The evolution of this organization from the academies of the seventeenth and eighteenth centuries to the universities and research institutes of the nineteenth and twentieth centuries, and of the

[32] Ralph H. Turner, "Role: Sociological Aspects," *International Encyclopedia of the Social Sciences,* Vol. 13 (New York: Macmillan and The Free Press of Glencoe, 1968), pp. 552–557.

scientific community from small groups and networks of intellectuals into large and powerful communities of professional scientists will be the subject matter of Chapters Six to Eight.

This last subject will be treated through three case studies of the development of the organization of science in France, Germany, and the United States. These chapters are arranged so as to explain the shift of the scientific center from Britain to France and, subsequently, to Germany and the United States. (The shift from Italy to England is discussed in the chapter dealing with the emergence of modern science.)

The reason for concentrating on the centers is that they have played a decisive role in the growth of scientific activity. This occurred because the questions how much and what kinds of scientific research there should be in a country have only in exceptional instances been decided on the basis of the social objectives to be attained through research. The reason for this is that even today there is no way of knowing precisely the relationship between different amounts and kinds of research (as distinct from *development*) and the attainment of various social objectives, such as the advancement of technology, economic progress, and military strength, that are believed to be the results of science.[33] Neither is there a satisfactory way of knowing the relationship between the choice of social structures (career patterns; definitions of roles; laboratory, departmental, university, and research organizations; national systems of training and research) for scientific work and different amounts and kinds of research.

The changing levels and forms of scientific activity through time and space have, therefore, developed through a kind of natural selection. In spite of recent attempts at the formulation of national science policies, actual developments up until now have been the result of disconnected initiatives and strategies adopted by those directly concerned with science, namely, the scientists, other intellectuals interested in science as cooperators with or competitors to scientists, and those who pay for science for their own or for some presumed public or private benefit. These groups have primarily scientific and/or more general intellectual purposes.[34] What they can do about them, however, is limited by their economic situation and political, religious, and other constraints. These constraints determine the social structures, such as the definition of the scientific role, and kinds of scientific organizations the groups can set up to realize their scientific goals. Usually they try to choose structures from available models and only rarely do they innovate new ones. They devise strategies toward the estab-

[33] For a further discussion of this problem, see Derek J. de Solla Price, "Is Technology Historically Independent of Science?" *Technology and Culture* (Fall 1965), VI:553–568; Schmookler, *op. cit.*; Ben-David, *Fundamental Research and the Universities*, pp. 55–61.

[34] About the sociology of intellectuals, see Theodor Geiger, *Aufgaben und Stellung der Intelligenz in der Gesellschaft* (Stuttgard: F. Enke, 1949); Logan Wilson, *The Academic Man, a Study in the Sociology of a Profession* (Fair Lawn, N.J.: Oxford University Press, 1942); Florian Znaniecki, *Social Role of the Man of Knowledge* (New York: Columbia University Press, 1940).

lishment of these social structures in the light of the existing constellation of forces in their respective societies.

Such a process can go wrong at any step. Economic interest, other constraints, or bad models combined with poor imagination may give rise to preferences for structures ill-suited to scientific research, and even if the structures are well chosen, the strategy may not succeed for one reason or another. The structures that survive do so in a fashion reminiscent of evolution. Where there is a good match between a structure and an ecology, the structure will thrive and spread.[35]

This process should give rise to a number of alternative social structures for the pursuit of science. But our evolutionary analogy has its limits. Although scientific roles are transplanted from one country to another, they do not lose touch with their place of origin. Science is communicated and learned over a network comprised of intellectuals who serve as role models, and those who carry science to far-away places copy the models. Thus links with the place of origin are not severed; the science transplanted to outlying areas forms a continuous periphery to that of the center. And the further development of the transplants is determined not only by their immediate environment, but also by the new environment of an international community that emerges as a result of the diffusion process.

Hence the evolutionary model must be modified. At every period, there are differences between the sociology of scientific work in the country (or countries) that are the temporary centers of science and the sociology of science elsewhere. In the centers—England during the latter part of the seventeenth century, France during the eighteenth century, Germany during the nineteenth century, and the United States at present—the social structures of science developed on the basis of the patterns of the previous center and innovations related to the conditions prevailing in the new center. Elsewhere, however, much of what happened was a response to, imitation of, resistance to, or competition with the center. Because of the unity of the world's scientific communities, members in the peripheral countries used the situation at the center as their frame of reference in thinking about their own conditions of work.

This way of viewing the sociology of intellectual activity is the rationale for treating the development of scientific roles and organizations as a process of diffusion and transplantation of models from one country to others. It also explains why we emphasize the succession of centers instead of systematically comparing the state of science in all countries.[36]

[35] For an exposition and example of the ecological approach, see Sir Eric Ashby and Mary Anderson, *Universities: British, Indian, African: A Study in the Ecology of Higher Education* (London: Weidenfeld and Nicolson, 1966).

[36] For the concept of centers, see Edward Shils, "Center and Periphery," in *The Logic of Personal Knowledge, Essays Presented to Michael Polanyi on His Seventieth Birthday, March 11, 1861* (London: Routledge & Kegan Paul, Ltd., 1961), pp. 117–130; Edward Shils, "The Implantation of Universities: Reflections on a Theme of Ashby," *Universities Quarterly* (March 1968), pp. 142–166.

19

As implied in the foregoing discussion, the succession of centers has to be studied from two different angles. One is the level of support for science. This may reflect the success of the scientists and/or their supporters in arousing interest in science among broad strata of the population. Such interest can lead to an increase in the motivation of young people to study science and to a diversion to science of part of the leisure-time activities of the wealthier classes. Until about 1830, direct interest in itself was a sufficient explanation of the ecology of scientific activity. The location of the center first in Britain and then in France was the direct result of the efforts of people who engaged in research spontaneously (Chapters five and six).

The other aspect of studying scientific centers is the adequacy of the organizations and systems of research. This has become an important determinant of scientific activity since the middle of the nineteenth century. Part of the problem belongs to the sociology of organizations and will be only touched on. But the strategies leading to the emergence or choice of systems, organizational forms, and role definitions in different societies; the working of the systems; and the effect of the system (as distinct from popular interest) on the level of scientific activity on the one hand and support to science on the other, are institutional topics and will form the main theme in the discussion of the German and United States centers (Chapters seven and eight).

the sociology of science

science in comparative perspective

two

The Lack of Continuous Scientific Growth
Before the Seventeenth Century

Rapid accumulation of knowledge, which has characterized the development of science since the seventeenth century, had never occurred before that time. The new kind of scientific activity emerged only in a few countries of Western Europe, and it was restricted to that small area for about two hundred years. Since the nineteenth century scientific knowledge has been assimilated by the rest of the world. This assimilation has not occurred through the incorporation of science into the cultures and institutions of the different societies. Instead, it has occurred through the diffusion of the patterns of scientific activity and scientific roles from Western Europe to the other parts of the world. The social role of the scientist (whether he is a university professor or a research worker in industrial or government laboratories) and the organizational surroundings of his work, in India, Japan, Israel, or the U.S.S.R. are varieties of social forms originating in Western Europe. They are not modifications of the traditional patterns of intellectual work that existed in these societies before the adoption of Western science. The question to be dealt with in this chapter is why the development of science beyond rudimentary beginnings was restricted to such a small fraction of human societies.

21

The Transmission and Diffusion of Science
in Traditional Societies

This lack of development cannot be explained either by the absence of the notion of science or by the absence of talent in the societies where science did not develop into a rapidly growing activity. Many societies, perhaps all, possessed a reasonably clear notion of the existence of necessary relationships between certain natural events and were capable of distinguishing this type of logical relationship from other types, such as magic and miracles. These societies created a certain amount of knowledge which might be considered as scientific.[1] And judging from the present-day performance of Chinese, Indian, Japanese, and other scientists, there must have been plenty of scientific talent in these societies. As a matter of fact, in some places, such as ancient Mesopotamia, Greece, and China, there were impressive achievements.

This impression that the retardation of science was caused by social, rather than hereditary, conditions or the absence of basic logical notions is further strengthened by the characteristic growth pattern of the scientific tradition in all societies before the seventeenth century. Relatively brief periods of flowering were followed by prolonged periods of stagnation and decline during which the scientific traditions actually tended to deteriorate. In the absence of a potential for scientific creativity there could have been no periods of flowering, brief or long. The recurrent phenomenon of decline has to be attributed, therefore, to deficiencies in the mechanisms of transmission and diffusion of the knowledge.

These deficiencies are evident from a comparison of the ways science is transmitted today with those prevailing in earlier times. Today there are journals, monographs, texts, and specialized courses of instruction. But in earlier times scientific knowledge was usually transmitted as part of the technological, religious, or general philosophical tradition. Thus most of the extant knowledge of ancient Egypt is found in religious or technical literature.[2] The same is true of the bulk of the Indian tradition. The Chinese tradition contains a number of technical treatises of a descriptive and classificatory kind, but theoretical works there were also part of philosophical and religious writings.[3] Something like specialized texts for the study of mathematics existed in the Babylonian and, in a more advanced form, in the Greek traditions. The latter also contained theoretical writings in other fields.[4] But even in this last case,

---

[1] Bronislaw Malinowski, "Magic, Science and Religion" in his similarly titled collection, *Magic, Science and Religion; and Other Essays* (Garden City, N.Y.: Doubleday, 1954), pp. 17–90.

[2] D. Guthrie, *A History of Medicine* (Philadelphia: J. B. Lippincott, 1946), p. 23; O. Neugebauer, *The Exact Sciences in the Antiquity*, 2nd ed. (New York: Harper Torch Books, 1957), p. 91.

[3] W. Brennand, *Hindu Astronomy* (London: C. Straker, 1896), pp. 133–134, 160; A. Rey, *La Science Orientale avant les Grecs—La Science dans l'Antiquité* (Paris: La Renaissance due Livre, 1930), p. 407; René Taton (ed.), *Ancient and Medieval Science* (London: Thames and Hudson, 1963), pp. 133–154; J. Needham, "Poverties and Triumphs of the Chinese Scientific Tradition," in A. C. Crombie (ed.), *Scientific Change* (London: Heinemann Educational Books, 1963), pp. 124–125.

[4] Neugebauer, *op. cit.*, pp. 97–190.

22

the independence of these scientific traditions from religious and metaphysical thought was restricted and ephemeral.

The most conspicuous example of how this packaging of the scientific in other traditions brought about deterioration is astronomy, by far the most developed science of antiquity. From its earliest beginnings, the tradition in this field had important astrological elements. Nevertheless, knowledge was based on observed astronomical appearances, and much of it was concerned with the practical problem of the intercalation of the lunar calendar. About the second century B.C., however, the center of attention turned to astrology of a magical kind, and this remained the principal preoccupation of the profession until the seventeenth century.[5]

Hence, even in the field where there existed a great deal of purely rational scientific literature, there were possibilities of deterioration through shifts of interest toward unscientific concerns.[6] Other sources of deterioration were the difficulty in preserving documents and mistakes made in copying manuscripts, especially during periods when a subject matter ceased to be of live interest.

Periods of scientific decline were usually much longer than those of scientific growth. They were relieved by "renaissance-like" phenomena. These, however, could not reestablish real continuity with the ancient knowledge, as that had usually been forgotten. Therefore, development had to start again, at times, from a lower level than had been attained in the past. The story of the deterioration and only partial rediscovery of the Greek tradition under the European Renaissance is too well known to need retelling. Similar developments also occurred in China. The ancient books had been destroyed at the end of the third century B.C. at the command of an upstart emperor, Shi Huang Ti. The destruction was part of an attempt to break the old feudal traditions, and in the Han period an attempt was made to restore them. In India, Buddhist supremacy had apparently disrupted the ancient astronomical traditions which were revived about 200 B.C., after the upset of that supremacy.[7]

This type of development suggests that there were several beginnings of creative scientific work in different societies. As a rule, however, these beginnings could not give rise to continuous scientific activity and, hence, to an accumulation of scientific knowledge. Sooner or later science was always subordinated to other concerns and, as a result, lost its vitality.

How did this subordination of science occur? Since the creation of scientific knowledge has always been the accomplishment of a very few people who

---

[5] *Ibid.*, pp. 168–171.

[6] Where scientific knowledge had not been separated from religious lore, such deterioration could result from mere changes in ritual. Thus, the oldest mathematical tradition in India is in the Vedic Sulva Sutras (the *Sutras* of the cord). This tradition did not influence the further development of geometry, and none of the geometrical constructions that were relevant to the old Vedic ritual occur in later Indian works. The ritual disappeared, and with it went a mathematical tradition. See W. S. Clark in G. T. Garret (ed.), *The Legacy of India* (Oxford: Clarendon Press, 1937), pp. 340–342.

[7] A. Pannekoek, *A History of Astronomy* (London: George Allen & Unwin, Ltd., 1961), p. 87; and Brennand, *op. cit.*, pp. 140–142.

were interested and competent in these matters, the way to answer this question is to determine what people were engaged in science in the earlier societies. This survey is likely to show us what their purposes concerning science were and why they were uninterested or unable to develop it further than they actually did.

The Social Roles of Contributors to
Scientific Knowledge in Traditional Societies

The people in traditional societies who possessed and created scientific knowledge were usually either technologists (including physicians) or philosophers. Therefore, in order to understand the transmission and growth of science, it is necessary to know what interests these various professional and intellectual groups had in the creation of a vigorous and self-transplanting scientific tradition. This chapter will attempt to do this for each of the relevant groups.

Engineers and other makers of tools and implements were usually humble people. Their names were preserved only when, in addition to their technological achievements, they were important as political or religious figures. Among people with technological or applied scientific interests, only those in astronomy, medicine, architecture, and construction were important enough to give their practitioners something like the professional status of today. In these fields there was, therefore, the possibility of developing strong intellectual traditions with some scientific content. The tradition which actually emerged, however, was not enough to give rise to continuous scientific activity for the following reasons:

(a) In all the technological traditions there was a discrepancy between the limited range of valid theory and the breadth of practical tasks. Perhaps the best example of this is astronomy-astrology, which has been mentioned above. Astronomical knowledge was suited for a limited set of practical tasks such as establishing a calendar, fixing the dates of seasonal festivals, and predicting heavenly events (e.g., eclipses), which were regarded as omens for various earthly events. All these tasks had been more or less satisfactorily mastered in Babylon, Egypt, Greece, India, and Mexico in quite early times. Within the limits set by the immediate practical purpose, there was no incentive for continued innovation.[8] Astrology, however, opened up unlimited tasks to the star-gazing profession, none of which were capable of scientific treatment. This restriction explains the instability of scientific creativity in the profession. In periods where astronomer-astrologists had scientifically feasible practical tasks

---

[8] In addition to Neugebauer, *op. cit.*, pp. 71–72; Pannekoek, *op. cit.*, pp. 87–90; J. H. Breasted, *A History of Egypt* (London: Hodder and Stoughton, 1906), p. 100; Taton, *op. cit.*, pp. 25–26; Brennand, *op. cit.*, pp. 25–26; and J. E. S. Thompson, *Rise and Fall of Maya Civilization* (Oklahoma: University of Oklahoma Press, 1954), pp. 160–164.

science in comparative perspective

(such as establishing a calendar, reforming it, or aiding navigation), there was a rise of scientific creativity. Once these tasks were completed, however, there were no more social incentives for creativity. On the other hand, high rewards were offered to astrological quacks at all times. Thus while genuine scientific search was stimulated only from time to time, astrological speculation was in permanent demand. Therefore, the scientific element could not become dominant in the role of the astrologer-astronomer.

(b) Discrepancy between limited scientific knowledge and an endless array of practical tasks also explains why medicine did not serve as a suitable vehicle for the preservation and improvement of a scientific tradition. Unlike astronomers, who had to answer a few valid and relatively easily answerable questions (as well as innumerable questions that had no valid answer at all), the questions put to physicians were all valid, but only very few of them were easily answerable. In principle this was not an unfavorable background for the development of a cumulative tradition of empirical and rational inquiry. The conditions that actually determined the behavior of the medical profession, however, were not the scientific possibilities inherent in the professional practice. Instead, medical behavior was determined by the requirements of healing and of making people believe in the ability of the doctor to help.

The practical task of healing has always tended to produce a number of physicians who were good observers and rational empiricists. At the same time, however, the need to make people (including themselves) believe in the effectiveness of medical help and the soundness of practice had exactly the opposite effect: it gave rise to a tendency of adopting general doctrines and impressive professional customs that were designed to create self-confidence on the part of the doctor and trust on the part of the patient.[9] Hence those in medicine vacillated between sober empiricism and unfounded theorizing.[10] This vacillation had a paradoxical result. Physicians had been the main source of a tradition of empirical inquiry and interest in natural science until the seventeenth (and to some extent even the nineteenth) century. These contributions to the sciences in general had little effect on medical practice or theory, however. The professional tradition in medicine had been conservative and doctrinaire. The profession as a whole displayed a great deal of caution and

---

[9] Thus there was a system of "Medical Ethics" in India. See J. Jolly, *Indian Medicine* (Poona: G. G. Kashika, 1951), p. 32. The Assyrian physicians of the seventh century B.C. used Sumerian formulas as later European physicians used Latin, and for the same reason. Sumerian was a nobler language and known only to an elite; its use conferred prestige on the doctor. See G. Sarton, *A History of Science*—Vol. 1, *Ancient Science Through the Golden Age of Greece* (Cambridge, Mass.: Harvard University Press, 1952), p. 89. About the traditionalism of Egyptian physicians see G. Foucart, "Disease and Medicine, Egypt," *Encyclopaedia of Religion and Ethics*, Vol. IV, pp. 751–752.

[10] A. Castiglioni, *A History of Medicine* (New York: Alfred A. Knopf, 1947), pp. 89, 94–95, for a description of Indian empiricism and magical theorizing; see also his discussion of Chinese medicine and "fanciful" theories, pp. 101–102. On unfounded theories on the heart, accompanied by well-developed empirical treatments in Egypt, see J. Pirenne, *Histoire de la Civilisation de l'Egypt Ancienne* (Paris: Editions A. Michel, 1961), pp. 198–204.

25

skepticism toward innovations, while it preserved and defended senseless traditions.[11] Thus, while medical work was an important source of individuals with a predisposition to engage in some kind of science, the medical community did not provide a social setting for the emergence of a scientific tradition systematically developed by the profession.

(c) The least problematic among the professional traditions had been that of the architects and construction engineers. Their tasks were as empirical as those of the physicians but much more limited and better defined. Architecture and engineering were also fields where magic and suggestion had little effect. Indeed, architecture and construction had remained the single entirely rational technological tradition pursued on a high intellectual level in several ancient and medieval civilizations.

But the rational purity of this technological tradition was not enough to serve as a basis for the emergence of science. In the long run, architecture and construction contributed much less to the growth of scientific knowledge than either astronomy or medicine, despite the fact that these latter seemed to have been hopelessly entangled with theology and magic on the one hand, and false doctrines on the other. The reason for this relatively meager contribution to science by architects and engineers was probably the circumstance that there was less need to express the architectural and engineering tradition in writing or in any abstract way involving the use of symbols. Medicine and astronomy dealt with phenomena that were only partly or not at all accessible to manipulation and close observation. Important parts of the model, by which the functioning of the human body or the movement of the heavenly bodies could be visualized and somehow grasped, were necessarily based on guesswork, and the guess had to be consistent logically. So there was need of some kind of theory.

On the other hand, the architect and engineer could see what they did and could handle their materials. Even if they used drawings, those represented concrete items or very simple abstractions like shape and distance, and not speculative models. Thus they could proceed and build structures or construct engines that were precise and usually much more complex than what could be grasped by available theory. Nor did they need a theory to establish their fame. That was proclaimed by the imposing structures they built that wore their name.[12]

[11] To a large extent this traditionalism was due to the risks that physicians ran of being accused of malpractice. See the quotation from the Greek historian Diodorus Siculus by Castiglioni:". . . For the physicians receive support from the community and they provide their services according to a written law compiled by many famous physicians of ancient times. And if after following the laws read from the sacred books they cannot serve the patient, they are let go free from all complaint, but if they act contrary to what was written they await condemnation to death, since the lawmaker thinks that few men would have knowledge better than the methods of treatment observed for a long time and prescribed by the best specialists." Castiglioni, *op. cit.*, p. 60. This is still the legal principle according to which malpractice suits are decided.

[12] About the relative absence of relationship between engineering and science see Neugebauer, *op. cit.*, pp. 71–72.

science in comparative perspective

(*d*) A common characteristic of all technologies has been their particularistic aim, namely the achievement of concrete result rather than the formulation of universal laws. This aim discourages accumulation and improvement of knowledge that comes from development of the logical implications of what is known, irrespective of their relevance to the immediate problem. This logical development maps out new areas for investigation and leads eventually to the discovery of contradictions between facts and theory and to the discovery of new theory. All of this development is lost if the aim of knowledge is merely to attain a particular practical purpose.

(*e*) The particularistic interests of technologists and other users of scientific knowledge may not only cut short further scientific inquiry, but may at times lead to opposition to any inquiry whatsoever. Since they are interested in providing a certain kind of service, their attitude to innovation will be determined by extrinsic considerations. For a priest who uses astronomy in order to fix the appropriate times for the various festivals of the year, there will be little incentive to adopt a standard calendar. His opposition may stem not only from a threat to his job, but also from such considerations as the loss of meaning of holy rituals. From a religious point of view such considerations are, of course, legitimate. Similarly, medical doctors, whose task is to heal people, may oppose on ethical grounds any preoccupation with theories and experimentation aimed at increasing knowledge if it has no direct effect on curing the sick.

(*f*) Finally, the technological purpose may lead to the forgetting of the scientific basis of the technologies even if scientific theory played a part in the original discovery. For example, if a technology is replaced by another, people may lose interest in the still valid scientific basis of the replaced technology.[13] Or, as another example, a technology may be perfected to such a level that its practice no longer requires the knowledge of the underlying scientific rationale (e.g. the introduction of calendars). This process happens every day when a science-based technological process is "developed" for manufacturing applications. At this stage all details of the process can be handled as well, or even better, by people entirely incapable of understanding its scientific basis rather than by scientists. In these cases, therefore, science embedded in technology tends to be forgotten in the transmission from the discoverer to the applier.

Philosophers as Contributors to Early Science

The second group within society that joins the technologists among creators of early science consists of philosophers. In their range of interests and aims, philosophers play the traditional role that comes nearest to that of the modern scholar and scientist. Not all philosophers have been interested in physical phenomena, but usually the person with the scientific

---

[13] See the case mentioned above in Note 6, which describes how discontinuing of the religious ritual that used specific mathematical knowledge led to the decline of that knowledge.

temperament who had a burning desire to understand nature in its own terms was most likely to be found among the philosophers.

Only a minority of traditional societies recognized that the philosopher maintained a role in his own right. Ordinarily philosophical inquiry was conducted by religious sages for whom philosophy was not an end in itself but a means to a way of life leading to salvation. Even if there were people amongst them whose personal aims were purely intellectual, those aims were either not understood or not accepted by the rest of society. In order to survive, philosophy had to be cast in a moral-religious tradition. The sage had to be a teacher and a model of the good life which was discovered by his philosophy. In actuality, the philosopher's role was also an applied rather than a pure intellectual one.

A good example of this is the story of the creation of the universe in the book of Genesis. The obvious intent of the story is to show that God created everything; that man is the crown of the creation; and that he was made in the image of God. For this purpose, however, a much simpler story would have sufficed, and one wonders whether the aim of the original story was not to explain creation in a way that made some logical sense. There are signs of struggle with the conceptualization of the state preceding creation, of an attempt to identify the original elements out of which matter and life arose, and to establish a view of geological and biological evolution. But in the form in which the story has been preserved, the original explanatory intent—if indeed there was such an intent—has been overlaid and obscured by moral-religious motives. The identity and the way of thinking of the natural philosopher who may have originated the story is entirely lost.

A similar fate befell Chinese and Greek natural philosophy. Eventually they became philosophical sects, often with a strong leaning toward magic and mysticism.[14] As a result, natural philosophy ceased to be a context of systematic inquiry altogether.

Natural philosophy survived as an intellectual pursuit subject to the laws of logic only in those few traditional societies where the social role of the general or moral philosopher developed. Aristotle was the outstanding figure and the model for this role type in the European and Near Eastern world for almost two thousand years. The Confucian scholar in China is a related variety, and there have been rudimentary versions of the same type of role elsewhere. This role appeared later in history than that of the technologist. Only in Greece and China was the role secular, and only in the Hellenistic world was it separated from the practical role of the lawyer and administrator. The social function of the philosopher was to find, through reasoning, a way that led to individual

[14] The promise of Taoism as a scientifically oriented natural philosophy was not fulfilled, because of its involvement with mysticism and its concern with everyday ethics as expressed by the *tao*—"the right way." See J. Needham, in Crombie (ed.), *op. cit.*, p. 134, and for a more detailed discussion see J. Needham, *Science and Civilization in China*, Vol. 2 (Cambridge: The University Press, 1954), pp. 33–164.

science in comparative perspective

and social perfection. The main emphasis was on metaphysics, and moral and political-legal philosophy (often accompanied by actual practice). The understanding of the place of Earth in the cosmos and of the place of man in nature was a subordinate part of their philosophical concerns.[15]

Since the philosopher is usually considered the direct predecessor of the scientist, it is important to see in what respect the two roles are similar or different. The traditional philosopher, like the scientist, is interested in grasping, by means of logical models, some kind of "reality." But the paradigmatic reality for the traditional philosopher was man and/or God. Natural events were not considered as important as human (or religious) affairs and were thought to be inaccessible to human reason and interference anyway. Reason was to be applied first and foremost to moral problems that it could grasp and solve for good purpose.

If roles are divided into means and ends, then there is a considerable similarity between the means of the traditional philosopher and the modern scientist: they both believe in logic and resort to empirical evidence. But the ends of the two roles are different: the philosopher wants to understand man intuitively in order to influence him, while the scientist tries to explain nature analytically in order to predict natural events. From the point of view of ends, therefore, the scientist is closer to the natural philosopher (even if the latter is a magician and a mystic), than to the general philosopher.

It is evident, therefore, that in the long run the social role of the general philosopher provided little incentive for scientific effort and creativity. However, it constituted the best framework for the preservation and occasional augmentation of the scientific tradition and (under certain conditions) for some real scientific work. The reason for this was that general philosophy, which was concerned primarily with the affairs of man and society, was rational and less likely to be influenced by magic and mysticism than the philosophy of nature, which wanted to penetrate the secrets of the universe. Common sense and practical knowledge of human affairs did not allow such flights of fancy as were characteristic of the speculations about life and death, the sun, the moon and the stars, or on lightning and thunder. Therefore, to the extent that scientific knowledge became part of the general philosophical tradition, it was more likely to be stated in the form of logical propositions than as knowledge which was transmitted in the traditions of the philosophy of nature. Furthermore, philosophers were schooled people possessing a mental discipline and often an unusual amount of curiosity, including, in some cases, genuine scientific curiosity. They were also likely to have more leisure than others. The likelihood, therefore, that they would occasionally contribute to science was considerable. Finally, as

[15] Ludwig Edelstein, "Motives and Incentives for Science in Antiquity," in Crombie (ed.), op. cit., p. 31; S. Sambursky, "Conceptual Developments and Modes of Explanation in Later Greek Scientific Thought," ibid., pp. 62–64; and "Commentary" by G. E. M. de Ste. Croix, ibid., pp. 82–84.

it has been pointed out, the philosophical purpose of providing men with a systematic cognitive orientation in his world did, in principle, include some attempts at the interpretation of the natural phenomena that baffled and troubled people in all ages (life, death, heaven, etc.).

Usually the incorporation of scientific elements into general philosophical systems led only to the falsification of scientific knowledge by forcing it into theoretical frameworks that did not suit it.[16] But at times, probably when new philosophical systems emerged and there was some openness and variety in philosophical thought, scientifically minded philosophers had an opportunity to emphasize the importance of their interests in natural phenomena and to devote systematic attention to them. It should be stressed, however, that within the accepted social definition of the role of the philosopher, this attention was not likely to lead to a takeoff into continuous scientific work. The aim of the philosopher, after all, was moral-social. In the long run, therefore, only those systems that offered a consistent solution to moral-social problems could become accepted by a significant part of the philosophical community. There was very little likelihood that the systematic study of nature should be an integral part of such systems.

Thus while the great questions about the motion of the stars, the origins of the universe, and the sense of fascination with the regularities of nature as perceived by the philosophers of nature were preserved in mystical and magical traditions, the logically formulated scientific theories became part and parcel of rational systems concerned primarily with metaphysics and moral philosophy. A continuous transmission of theoretical knowledge could be expected only in the framework of this general philosophy. But the intent of this rationality was nonscientific, and there was very little in it to inspire scientific creation. Creative scientific genius was therefore more likely to be attracted to the mystical-magical tradition than to rational general philosophy. However false its answers, the mystical tradition was more likely to preserve the great inspiring questions of natural inquiry. From the point of view of the moral philosopher or the legal expert, such questions seemed like useless and crazy things to pursue.

This attitude explains the unexpected connection between magic and mysticism on the one hand, and exact natural science on the other, which has been observed in a number of instances. Neo-Platonism, for instance, had been widespread among sixteenth and seventeenth century scientists. Kepler and (in some ways) even Newton were mystics as were the ancient Pythagoreans, the depositories of one of the earliest scientific traditions.[17] Lately, there have

---

[16] J. Needham, *Science and Civilization in China,* Vol. 3, pp. 196–197, for a Taoist criticism of the Confucian scholars for "forcing facts and going against Nature." The same criticism was applied in the seventeenth century against scholastic philosophy in Europe.

[17] J. M. Keynes, "Newton, the Man," in J. R. Newman, *The World of Mathematics,* Vol. 1 (New York: Simon & Schuster, 1956), pp. 277–285; L. Thorndike, *A History of Magic and Experimental Science,* Vol. 7 (New York: Columbia University Press, 1958), pp. 11–32; *ibid.,* Vol. 8, pp. 588–604; A. Koestler, *The Sleep-walkers* (New York: The Macmillan Company, 1959), pp. 261–267 (on Kepler) and pp. 26–41 (on Pythagoras).

been suggestions that Faraday and Oersted were influenced by the romantic philosophy of nature.[18]

On the surface, therefore, the scientific tradition survived in the form of the sterile but logically argued theories, or in the refreshing but untheoretically technical treatises of practical people trained in rational philosophy. But there had been a kind of subterranean link between science and magic and mysticism; that link manifested itself in unexpected eruptions of scientific creativity.

Conclusion

The slow and irregular patterns of scientific growth described at the beginning of this chapter can be explained, therefore, by the absence of the specialized role of the scientist and the nonacceptance of science as a social goal in its own right. Roles are initiated by people interested in expressing themselves in a certain way. But in order to become accepted by others and perpetuated, people have to fulfill a recognized social function. Technology is a necessary social function everywhere. But, as has been shown, only under very special conditions do technologists have more than a cursory interest in science. The same applies to general philosophers. Although finding the truth in general is an essential part of the role of the philosopher, the main purpose of this intellectual role in the past was to help people deal with their uncontrollable anxieties, and to curb their passions and harness their energies to the practical tasks of survival and betterment of the individual and his society. Finding the "truth" to philosophers meant spiritual and moral truth, not any sort of scientific truth.

Before science could become institutionalized, there had to emerge a view that scientific knowledge for its own sake was good for society in the same sense as moral philosophy was. Something like this idea had apparently occurred to some natural philosophers. But in order to convince others that this was so, they had to show some moral, religious, or magical relevance of their insights. As a result, the scientific content of natural philosophy was either lost or concealed by the superstitions and rituals of esoteric cults.

These reasons explain the absence of a significant growth of science in the overwhelming majority of civilizations. Science, unlike technological services, moral philosophy, or religion, was something that was not "needed." Even the purely intellectual interest in understanding man's place in nature could be satisfied only by general philosophies and not by testable scientific enquiry.

Thus the initial question of why science did not grow more rapidly and did not bring about the passing of traditional societies earlier and in more places than actually occurred, has been answered. Actually, the question can be turned around. What needs explanation is the fact that science ever emerged at all. Students of traditional societies may argue that there is something pathological

[18] See Pierce Williams, *Michael Faraday* (London: Chapman & Hall, Ltd., 1963).

in the rapid growth of science which has occurred in the West. They may argue further that the pattern of a slow and intermittent growth of scientific knowledge closely linked to technology and moral philosophy, which is characteristic of traditional cultures, represents a more balanced social and cultural growth than our own.[19]

[19] For such an argument see J. Needham, in R. Dawson (ed.), *The Legacy of China* (Oxford: The Clarendon Press, 1964), pp. 305–306.

the sociology of
greek science
three

Greek Science as A Forerunner of Modern Science

The principal conclusion in the previous chapter was that the emergence of a distinct scientific role and continuous scientific activity depended on social conditions that did not exist in any of the traditional societies. The single case that apparently contradicts this view is that of Greek science which can be considered, from the point of view of its logical structure, as the legitimate ancestor of modern science. Therefore, before it is possible to identify the social conditions that led to the emergence of modern science, it is necessary to see to what extent and in what respects Greek science can be regarded as a genuine forerunner of modern science in the sociological sense. If the development of Greek science was indeed such that the rise of modern science in the seventeenth century can be justified as a simple extension of the Greek trend (temporarily interrupted by the rise of Christianity and barbarian invasions), then our description of the rationale of traditional science was wrong. The search for social conditions of the emergence of modern science in the sixteenth and seventeenth centuries will then be misplaced, since the conditions of the origins of the scientific role should be sought in ancient Greece.

In order to investigate this question the following information is necessary:

1. What were the social roles of the people who contributed to Greek science? Were they recognized and honored as scientists, or was science only a

secondary activity for them? Was science considered as important in its own right, or only because of its broader philosophical or mystical implications, or its technological applications? And if those who contributed to science were considered first and foremost as scientists, was the definition of this role similar to its definition today, namely a person seriously engaged in the augmentation of scientific knowledge? Or was it one of the universally existent marginal and semi-institutionalized roles, such as that of the expert in a rare but seldom-needed skill, such as the chess champion, the person with a phenomenal memory or an extraordinary ability for mental calculations, and so on, who is admired or, at least, talked about as most unusual beings are.

2. What were the uses made of science? Was it systematically taught as an important part of the culture in its own right? Or was it considered only as an auxiliary study to another subject (philosophy, ritual, technology) or as a mere intellectual pastime?

3. What were the patterns of the transmission of scientific knowledge? Were there institutions designed to spread and improve such knowledge among the educated circles of the population? Was science transmitted like the secret lore of a cult or like a trade secret from master to disciple? Or was it transmitted in a random variety of ways, as matters of only occasional interest and of no great importance tend to be?

4. What was the resulting pattern of scientific growth? Was it cumulative or the oscillating kind with periods of decline following periods of rise? Finally, were the ups and downs of scientific activity determined by changes in the internal state of science or by the changing uses and meanings of science in a variety of contexts external to scientific knowledge?

The answers to the first and the second pairs of questions are interrelated. The role of the scientist is not defined independently of the uses made of scientific knowledge by himself and others. There is also a close relationship between the ways scientific knowledge is transferred and spread through time and space and the patterns of its growth. Therefore, the information will be presented in a chronological order, rather than as separate answers to each question.

The Social Roles of Early Philosophers of Nature

The social roles of the pre-Socratic Greek thinkers were similar to those of the great intellectual figures of other traditional societies. Some of the early philosophers, like Pythagoras or Empedocles, were saintly men, founders of religious cults who were honored as prophets in their lifetime. Their mythical biographies remind one of Buddha and Zoroaster, or of the Jewish prophets. Giorgio de Santillana has compared the story of how Pythagoras lay for a month in a cave in Crete as part of his initiation with the similar story

34

of how Ezekiel lay stretched out for 390 days as a God-imposed penance.[1] Another possible comparison is the ascendance of Elijah to heaven in a burning chariot with the end of Empedocles (another famous healer) in the Etna volcano and with the Pythagorean idea of the ascendance of the soul to heaven. Other thinkers—and again there is no lack of parallels with ancient Jewry and many other tribes even in modern times—were preachers who castigated the immoral customs of the rich and turned against ritual traditions and priestly caste in the name of a true, more rational, morality. Finally, in all these cases there emerged groups of masters and disciples who set themselves apart from the rest of society and cultivated religious and moral practices of their own. Ultimately, these practices took the shape of religious brotherhoods devoted to a cult, such as the Pythagoreans' cult of the Muses and of the apotheosized Pythagoras. Besides Pythagoras himself and many of the rest of the early philosophers were leaders and lawgivers in the outlying Greek colonies who were continuously fighting for their survival, which again recalls the figures of Moses and Ezra.[2]

It is quite possible that there were exceptions which do not fit into this typology. Anaxagoras and perhaps one or two others appear to have been more secular and more specialized philosophers in the modern sense.[3] But whether this impression is correct or not, the fact remains that to the extent that there developed a continuous activity and tradition, this took the form of cults concerned with knowledge and truth in the ultimate religious sense and with their expression as a way of life. The special role of the natural philosophers in Greece was therefore similar to those of other ancient civilizations. The only difference seems to have been in the better preservation of the personalities and the teachings of these thinkers, including those whose teachings were clearly heretical.

This preservation might have been due either to the better quality of the people and their thinking, to special conditions, or, most likely, to the interaction of both. Where conditions are more favorable to a certain kind of activity, better people are attracted to it. These conditions apparently existed in the case of the Greeks. They were a very dispersed and politically decentralized nation possessing a common culture and religious center. The only parallel which comes to mind, incomplete though it may be, is that of the English-speaking nations today or the German-speaking nations in the first half of the nineteenth century. In centralized, homogeneous cultures there was no room for several different prophecies, or for cults potentially dangerous to religion. If there were differences, they led to clashes in which only one party survived.

[1] Giorgio de Santillana, *Origins of Scientific Thought* (New York: Mentor Books, 1961), p. 55.
[2] *Ibid.*, pp. 54–57; Edward Zeller, *Outlines of the History of Greek Philosophy* (Cambridge: Cambridge University Press, 1957), pp. 216–217.
[3] Daniel E. Gershenson and Daniel Greenberg. *Anaxagoras and the Birth of Physics* (New York: Blaisdell, 1964); Zeller, *op. cit.*, pp. 76–80.

The Greeks, however, could produce diversity. What was dangerous in Athens or Sparta could be admissible in Miletus or Syracuse. They had a "frontier" in the sociological sense. Bands of adventurers and dissenters could move out and experiment with new political and religious ideas in small, selected circles without colliding head on with established religion and government and without cutting themselves off from the nation and its culture. This frontier made possible much bolder flights of imagination and experimentation with religiously and politically incomplete or even dangerous notions. Such notions gave rise to scientifically modern and almost specialized secular ideas that do not have many parallels in other traditional civilizations.[4] Some of these ideas make one feel that the takeoff into science must have been around the corner, and one becomes puzzled why the takeoff did not actually take place. However, if the situation is considered from a sociological point of view, it becomes evident that this intellectual activity did not constitute the beginnings of a socially accepted scientific role. Whatever motivations certain individuals might have had, the actual social purpose of the pre-Socratic groups was either to find the way to good life through special ways of understanding nature (which were usually only suitable to esoteric sects), or to combine people with particular skills and interests in some kind of guildlike societies.

Science in the Great Schools of Philosophy in Athens

This situation changed in the period following the Persian wars. Greece had become an increasingly unified state. The fragmentation of the country into independent city states that were closed religious communities gave place to coalitions, overlordships, and eventually to incorporation in the Macedonian Empire. Increased social mobility and a more complex structure of social classes and political forces gave rise to problems of government and public and private morality for which the traditional ways did not provide an answer.

Thus a need for the two specialized intellectual roles of the kind to be found in all traditional societies arose: lawyers-administrators-politicians on the one hand, and religious-moral-philosophers on the other.[5]

This need created a demand for the teaching of philosophy and intellectual techniques in general. The social demand for people to perform these tasks provided new opportunities for the intellectually alert and for trained members of philosophical cults. To remain esoteric and aloof must have been as difficult for them under the new conditions as it was for the Puritan artisans and small merchants of the seventeenth and eighteenth centuries to remain poor and pure by refusing to profit from the opportunities of expanding business enterprise and industry.

[4] For the concept of "the frontier" as used here see C. E. Ayres, *Theory of Economic Progress* (Chapel Hill, N.C.: University of North Carolina Press, 1944), pp. 132–137.
[5] Zeller, *op. cit.*, pp. 112–114.

the sociology of greek science

The need for trained people led to the rapid rise of schools to train people in the skills needed for public affairs—argument, effective thinking, and oration. Eventually there arose the roles of the orator and the lawyer.[6]

Of greater interest in the present context is the role of the philosopher. The practically minded schools of oratory used many of the intellectual techniques developed by the philosophers, including some of the scientific techniques. But their practical purpose could not satisfy the philosophers whose aim was to acquire valid knowledge. Nor was it in the power of oratory and effective public argument to solve the moral and political problems of the rapidly changing Greek society.

This situation gave rise to a new type of philosophical and scientific interest. Public teaching of philosophy and application of intellectual techniques brought members of philosophical-scientific sects out of seclusion. Their esoteric doctrines were confronted with others, and their usefulness in explaining things or guiding human action was openly examined.

The intellectual debate between schools and doctrines led to rapid advances in thinking. Within a generation after Socrates, who marks the onset of this new nonsectarian philosophical quest, philosophy had become abstract and professional. The predominance of moral-religious and practical problems led to an emphasis on these and on metaphysics. There was an obvious tension or even conflict between the philosophical purpose of arriving at a universally valid approach to knowledge that would supersede the doctrines of philosophical sects, and the social purpose to provide a new foundation for morals, religion, and politics. A resolution to the conflict between these two purposes was necessary because of the rapidly disintegrating ways of tradition. Philosophically, the most important challenge was the logical and scientific achievement of the natural philosophers. But logic and science were the least relevant for the solution of moral-religious and political problems. The burning question was how to make people and governments good at a time when traditional morality ceased to be an effective guide of private and public conduct.

This tension led to attempts of reinterpretations and assimilation of the philosphies of nature into new metaphysical systems. The attempts took place amidst actual encounters and debates between the representatives of the new schools and members of scientific sects and guilds. Plato had close associations with Eudoxus of Cyzicus, and the latter took interest in the philosophical discussions of the Academy. Plato's travels also brought him to the Pythagoreans at Tarentum and the medical school of Philistion in Sicily.[7]

These associations became actual collaboration in the Lyceum founded by Aristotle. Theophrastus, Eudemus, and Menon were charged with the writing of the histories of different disciplines. Close connections were maintained be-

[6] H. I. Marrou, A History of Education in Antiquity (New York: Sheed & Ward, 1956), pp. 189–312.
[7] Werner Jaeger, Aristotle: Fundamentals of the History of His Development (Oxford: Clarendon Press, 1948), pp. 17–18.

tween the Lyceum and the Cnidian school of medicine, and actual scientific investigations in zoology and anatomy were carried out at the school.[8]

The influence of these connections was quite different on Plato's philosophy than on that of Aristotle. Both men related themselves explicitly to the old philosophies of nature and concluded that their attempts to understand the physical universe in its own terms were too crude and logically unsatisfactory. They then resolved the difficulty by shifting the emphasis from physical to human problems. Plato accomplished this through the rejection of observations in favor of a world of ideas. The watching of stars and the geometric representation of their motions was a meaningful pursuit in Platonic philosophy. Its purpose, however, was to make "the mind look upwards . . . to higher things" (Glaukon). If so, one could also conclude with Socrates that since the native intelligence of the soul was the final source of higher knowledge, one could as well leave the starry heavens alone. And from this point of view, Pythagorean experiments with actual chords and sounds seemed entirely useless.[9]

For Aristotle, mathematics was less important as a subject than empirical science. Much of his own work, as well as that of his followers, was concerned with empirical problems of physics, astronomy, zoology, and botany. But the tradition that grew out of his work lost its scientific character after two generations and became a dogmatic system that subsumed all knowledge under a philosophy in which the anthropomorphic concept of "purpose" served as the basic principle.[10] A shift from natural to moral philosophy took place in the atomistic school. Epicurus had to choose between the belief in man's freedom of will and the rule of necessity in the universe as formulated by Leucippus and Democritus. He chose freedom of will and thereby cut off a potentially powerful line of inquiry into the laws of nature.[11]

The end result of these developments appears to have been a sociological parallel to what happened in China. Confucian philosophy also had the intent of a codification of all knowledge within the framework of a political-moral philosophy, as did the later philosophical schools in Greece. Similarly, there arose in both places a role of the philosopher as the guide to the political leader and, ideally, as the political leader himself.[12]

The differences were variations within the same sociological role type, if role type is defined by its goals and the means applied towards attainment of those goals: the purpose of philosophy is to create a good society where the wise are the leaders and there is a common assumption that the way to the attainment of a perfect society can and *has been* actually discovered by a very great philosopher. The means to attain the desired end are, therefore, the proper study and

[8] *Ibid.*, pp. 335–337.
[9] S. Sambursky, *Physical World of the Greeks* (London: Routledge & Kegan Paul, Ltd., 1962), p. 54. For the way Plato used the various sciences for his own philosophical purpose see Jaeger, *op. cit.*, pp. 17–18.
[10] Jaeger, *op. cit.*, p. 404.
[11] Sambursky, *op. cit.*, pp. 109, 161–165.
[12] Marrou, *op. cit.*, pp. 47–54, 63–65, 85–86.

38

the sociology of greek science

practice of philosophical principles, especially by those who are destined to lead society. The Chinese had been more successful in attaining the ideal of a society run according to philosophical principles that were actually administered, to some extent, by philosophers themselves. Nevertheless, in both civilizations, there developed some differentiation between the person who taught philosophy and the one who practiced it in politics.[13]

In the same period, however, there emerged intellectuals such as Aristarchus, Eratosthenes, Hipparchus, Euclides, Archimedes, and Appolonius whose work can be considered as specialized and professional science. This development did not have a parallel in other societies. It therefore poses the crucial question of the extent to which this development represented the first emergence of socially recognized scientific roles.

The Separation of Science from Philosophy in the Hellenistic Period

In order to reject the suggestion that these were scientific roles in the modern sense, one has to accept the interpretation which goes furthest in attributing a modern character to this development.[14] According to this interpretation, Aristotle represented an important turning point. His system —as it existed in his own time, and not as reinterpreted by the later Peripatetics —was a relatively open one. Although it had an ultimate metaphysical-religious purpose (to prove that a supernatural reality existed) it was to a large extent a method of inquiry that allowed a great deal of autonomy to the specialized disciplines. Accordingly, there was a genuine division of labor in the Lyceum between specialists pursuing different substantive fields, but they were united by a common philosophical purpose. Had this philosophical purpose been accepted as a social value, and had it been consistent with a genuine freedom of disciplines, then a scientific role could have arisen that would have been recognized as a social function.

As it turned out, however, the philosophy did not completely satisfy either educated people in general or scientific specialists. The former needed a world view relevant to their religious and moral problems, and they were not very interested in empirical science. The scientific specialists needed freedom from any ultimate metaphysical purpose. Physicists and mathematicians in particular could not fit their methods into the teleological framework of Aristotelian metaphysics. As a result, the pursuit of science became a peripheral concern in the schools of philosophy and was conducted in more specialized circles.[15]

[13] Tilemann Grimm, *Erziehung und Politik im konfuzianischen China der Ming Zeit (1368–1644)* (Hamburg: Mitteilungen der Gesellschaft für Natur and Völkerkunde Ostasiens (OAG) Band XXXV B, 1960, Kommissionsverlag Otto Harrassowitz, Wiesbaden), pp. 33–35, 108–111, 120–121, 129–130, 159.

[14] This is the interpretation of Jaeger, *op. cit.*, pp. 335–341.

[15] *Ibid.*, p. 404.

This development, which may seem to be the beginning of the scientific role with a socially recognized purpose and dignity of its own, was, as a matter of fact, a sign of failure. The newly differentiated role was never given a dignity comparable to that of the moral philosopher. Independence from philosophy was a decline and not a rise in the status of the scientists. During the period when Plato and Aristotle tried to recast the moral-religious foundations of Greek society and of the intellectual foundations of Greek thought, science was drawn into the center of the intellectual concerns of society. Mathematics and physics, even if misrepresented, had the same moral significance as ethics in the ideal of the theoretic life of Plato.[16] With Aristotle, this tendency was to some extent even reinforced.[17] But starting in the third century, the few astronomers, mathematicians, natural historians, and geographers who worked mainly in Alexandria were completely isolated from any general intellectual or educational movement. They were advisers on military matters (Archimedes), astrologers, or simply parts of the entourage of the court. Thus it was increasingly difficult to maintain the postulated unity and ethical significance of all scientific and philosophical endeavor. Consequently, in the generation following that of Aristotle, one of the members of the Peripatetic school, Dicaearchus of Messene, declared the irrelevance of pure theory for ethics and practical life.

According to this view, specialized science (which included metaphysics) lost its moral importance. Even in later philosophies, when the idea that theoretic life was the only way to the supreme good was renewed, this acknowledgment did not refer to science, but to religious contemplation and metaphysics.

The Special Case of Greek Science in a Traditional Social Structure

Thus if attention is focused on the scientific role and on scientific activity, a different picture emerges from the one obtained from observation of the substantive development of science. From the substantive point of view the development may easily appear as linear and cumulative: the confused traditions of natural philosophies of the sixth and fifth centuries B.C. became systematic and logical during the fourth century and specialized and technical by the third.

However, a review from the point of view of the development of the

[16] O. Neugebauer, *The Exact Sciences in the Antiquity*, 2nd edition (New York: Harper Torch Books, 1957), p. 152; G. E. M. de Ste. Croix, in A. C. Crombie (ed.), *Scientific Change* (London: William Heinemann Ltd., 1963), pp. 82–83.

[17] The presentation of these changes in the moral education of the theoretic life from Plato to Dicaearchus is based on Jaeger, *op. cit.*, pp. 426–461. See also Zeller, *op. cit.*, pp. 136–137; R. Taton (ed.), *History of Science*, Vol. I: *Ancient and Medieval Science from the Beginning to 1450* (New York: Basic Books, Inc., 1963), Chap. 5, pp. 248, 262–265; Paul Friedlander, *Plato: An Introduction* (New York: Harper & Row, 1958), pp. 85–107; A. E. Taylor, *Aristotle* (London: T. C. & E. Jack, 1919), pp. 14–35; G. Sarton, *A History of Science*, Vol. I: *Ancient Science Through the Golden Age of Greece* (Cambridge, Mass.: Harvard University Press, 1952), pp. 331–347.

the sociology of greek science

scientific role and scientific activity produces a different picture. Early beginnings, which took place on the frontiers of Greek society by marginal groups, were transferred during the fourth century to the cultural and political center of the society as part of a comprehensive philosophical program that had moral and religious intents. As long as this program was in a state of flux, its theoretical structure unfinished, and its intellectual and practical scope unexplored, science was considered as an integral part of this enterprise. This program stimulated science, influenced its course, and provided it with a universal moral dignity that science did not have before. As soon as the philosophical enterprise became stabilized, however, it could accord only a very peripheral role to science. At this point science became differentiated from philosophy. The new autonomy did not, however, confer greater dignity on the scientists. On the contrary, it made evident the marginality of their concerns. As a result, the role did not develop any further and, starting from the second century B.C., scientific activity declined.[18]

The principal evidence in favor of this interpretation is the relatively short creative period of Greek science, which is otherwise impossible to explain. In medicine and astronomy, which were applied subjects, the creative period lasted until the end of the second century A.D.[19] But in the pure sciences of mathematics and physics, creativity came to an end about 200 B.C. Had the autonomy of specialized science enhanced the dignity of the scientific role and the motivation to engage in research, one would have expected an acceleration of scientific creativity in the second century B.C. This was just about the time when the breakup of the Aristotelian integration of science with philosophy became final as a result of the rise of philosophical-religious systems (especially the Stoa) and the changes within the Peripatetic school itself.

Further evidence is provided by the just-described development of the view about the moral significance of theoretic life. Had the differentiation of science been a step toward the public recognition of the importance and dignity of this role, then there would have had to have been an attempt at the creation of an ideology claiming a dignity in its own right that would have been for specialized science equal to but independent from philosophy. There ought to have arisen some kind of ideology about the specificity and superiority of the scientific as compared with the philosophical way to knowledge. The rise of such ideologies marked the rise of modern science in the seventeenth century and, again, the emancipation of German natural science from philosophy after 1830. However, there was nothing of this sort in Greece. Concern with matters of intellectual dignity seems to have been prevalent among members of the philosophical schools, and of the brotherhoods of physicians. But these claims contained nothing about

[18] S. Sambursky, "Conceptual Developments and Modes of Explanation in Later Greek Scientific Thought," in A. C. Crombie (ed.), *op. cit.*, pp. 61–63.

[19] Ludwig Edelstein, "Motives and Incentives for Science in Antiquity" in Crombie (ed.), *op. cit.*, pp. 25–37, and "Recent Trends in the Interpretation of Ancient Science," *Journal of History of Ideas* (October 1952), XIII:573–604.

the specific dignity of the scientist as distinct from that of the philosopher. To the contrary, the philosophical unity of all intellectual endeavor was emphasized.[20] But as has been shown above (pp. 39–40), the scientists' claim to be considered an important part of the philosophical enterprise was denied to them after the passing of the heydays of the Aristotelian experiment. Hence the beginnings of specialized science in the Alexandrian period did not augur the rise of a distinct scientific role and the acceleration of scientific work. The status of science, its educational or other public influence remained extremely limited throughout the entire Hellenistic period. The principal philosophical schools either paid lip service to science or were actually hostile toward it. Science was virtually absent from the curricula of the rhetorical schools which were the most widespread educational institutions.[21]

Conclusion

Thus even if it could be shown that a few individuals developed a self-image comparable to that of the modern scientists (which is doubtful), this would still not prove that there had emerged a publicly recognized role of the "scientist," which was different from that prevailing in other traditional societies. Irrespective of any individual inclinations or personal scales of preference, the Greek public viewed the scientist either as a philosopher, or, if as a specialist, as a person with a peculiar interest of no great social significance.

The rhythm of scientific creativity was also similar to that of other traditional societies. The periods of flowering were relatively brief occurrences. As it has been pointed out, the main developments occurred within a span of less than two hundred years, between the fourth and the second centuries B.C. This was a time of basic philosophical transformation, when a new world view was to be established and all knowledge had to be reviewed and systematized. As far as the general, or the philosophical publics were concerned, science was only a means to the end of systematization. Once this was accomplished, interest returned to the practical and socially useful aspects of philosophy. Science was simply filed in its appropriate place in one system or another. This growth pattern is common to all the traditional societies. The great spurt of scientific creativity was due to events external to science (the emergence of new philosophical world views), and not to immanently scientific occurrences, such as great discoveries that stimulate new developments. And when the external conditions changed, stagnation followed. As soon as science was cut off from the moral concerns of philosophy, there was no scientific community stimulated by its own problems and capable of convincing society in general about the importance of its enterprise.

This explanation is not contradicted by the rise of Ptolemaian astronomy at the end of the second century. As it has been pointed out, in astronomy-

[20] Edelstein, *op. cit.*, pp. 33–41.
[21] Marrou, *op. cit.*, pp. 170, 189–191. Edelstein, *op. cit.*, pp. 31–32.

the sociology of greek science

astrology and medicine, the period of stagnation set in only after the second century A.D. when the actual attacks on science began. These were practical subjects, nurtured in all traditional societies, so that their development did not so much depend on their relationship to philosophy. The only theoretical advance of great significance that occurred in one of these fields, the rise of Ptolemaian astronomy, was not due to intellectual developments within the astronomical community but to the fusion of Babylonian with Greek astronomy.[22] Both the concentration on applied fields, and the theoretical advance occurring in such a field (not as a result of the immanent development of research, but of culture contact), are characteristics of traditional science. The greater continuity of medical and astronomical research and the Ptolemaian revolution in astronomy, do not, therefore, contradict the interpretation that about 200 B.C. Greek science settled back to the traditional pattern but rather confirm it.

In conclusion it has to be emphasized that the classification of the Greek case as "traditional" in its social structure only explains the absence there of (a) a socially recognized and respected role of scientist, and (b) a scientific community that could set its own goals relatively independently from nonscientific affairs. This examination explains the growth pattern of Greek science which, although it rose much higher than the science of any other traditional society, had a shape similar to that of the rest of societies in the same category: a brief and brilliant period of flowering, preceded by an incubatory period and followed by prolonged stagnation and eventual decay.

This overall view does not deny the intellectual excellence of Greek achievements. Part of this excellence was due to sociological conditions. As previously explained, the existence of a Greek frontier during the incubatory period of science (not wholly eliminated even under the Hellenistic and Roman empires), created superior conditions for secularism, cultural contacts, and cultural pluralism. This frontier gave rise to a greater variety of natural philosophies, lent them greater depth, and allowed a greater differentiation of intellectual roles than elsewhere.

The second unique characteristic of Greek science was the great importance attached to mathematics. This is an unusual feature in traditional science but not inconsistent with it. There is no way to explain how this mathematical importance happened. It probably had to do with the connection of mathematics with music. Music had been a very important element in ancient Greek culture and education (as it had been in several other places).[23] The discovery of the ratios expressing the concordant intervals in music by Pythagoras had been a stroke of genius, but such discoveries had also occurred elsewhere.[24] The connection between this discovery and the harmony of the universe had been

[22] Derek J. de Solla Price, *Science Since Babylon* (New Haven: Yale University Press, 1961), pp. 14–17.
[23] Marrou, *op. cit.*, pp. 55–56.
[24] Taton, *op. cit.*, Vol. 1, pp. 168–169, 215–216, 438.

43

a nonscientific speculation of the kind characteristic of natural philosophy everywhere. But the combination of these occurrences, all of which had been within the range of variation generally characteristic of ancient scientific thought, was enough to make mathematics part of one of the mainstreams of Greek thought for a fairly long period of time.

The technical nature of mathematics made it more difficult to assimilate the scientific elements into the comprehensive philosophical and religious world views and the technological traditions in Greece than elsewhere. The tension between the metaphysical and moral elements on the one hand and the scientific ones on the other could never be eliminated from the Aristotelian tradition and was, at least potentially, present in Platonism too.[25] This tension persisted throughout the Middle Ages and became manifest in the sixteenth and seventeenth centuries at the time of the rise of modern science. The discourse that arose then could be directly related to the Greek tradition, which in this respect has a unique importance for the rise of modern science. But it must not be forgotten that the intellectual tension built into the Greek tradition, like any cultural tension in itself, did not and could not create the social recognition, means, and motivation necessary for the emergence of continuous scientific activity. For this, widespread social interests were needed. The emergence of these social interests in the seventeenth century will be the subject matter of the next chapter.

[25] This was much less the case in the most popular philosophies of late antiquity, the Stoa and Neoplatonism; see also de Ste. Croix, *op. cit.*, p. 81; Sambursky, in Crombie (ed.), *op. cit.*, p. 64; Edelstein, in Crombie (ed.), *op. cit.*, pp. 31–32; Marrou, *op. cit.*, pp. 120, 189–191.

the emergence of
the scientific role
four

According to the conclusions of the previous chapter ancient science failed to develop not because of its immanent shortcomings, but because those who did scientific work did not see themselves as scientists. Instead, they regarded themselves primarily as philosophers, medical practitioners, or astrologers. It is true that much of the Greek tradition was lost in the Middle Ages as a result of wars and vandalism, but the stagnation and deterioration of the tradition had started earlier. Furthermore, had there been a group of persons who inherited the Greek scientific tradition and regarded themselves as scientists anywhere in the Christian or Moslem world or among the Jews, the Greek achievements might have been rediscovered in the Middle Ages. If either had been the case, much more efficient use would have been made of the Greek achievements in the fifteenth century when they were rediscovered.[1]

The question, therefore, is what made certain men in seventeenth century Europe view themselves for the first time in history as scientists and see the scientific role as one with unique and special obligations and possibilities. What made this self-definition socially acceptable and respectable? The explanation will be of the evolutionary type: the new role emerged as the result of several independent developments that were eventually incorporated into the role of the scientist.

[1] George Sarton, *The Appreciation of Ancient and Medieval Science During the Renaissance* (Bloomington: The University of Indiana Press, 1957).

The Emergence of the Professional
University Teacher
in the Medieval University

In traditional societies the typical form of higher education was found in a master surrounded by disciples. Some of the disciples might become quite famous scholars in the lifetime of the masters, but only one could inherit the master's position. The other disciples might establish schools of their own elsewhere to carry on their masters' traditions, or they might inherit the leadership of an existing school whose master did not leave a disciple worthy or capable of inheriting from him. Rulers, rich individuals, or a community usually supported such a school by granting privileges to the scholars and maintaining hostels and teaching halls, paying the master a salary or gifts, or establishing an endowment.

Masters and scholars might be motivated by a genuine desire to understand sacred truths, to gain honor, or by anything else. The legitimation of learning, however, was that it was "practical"; it prepared the pupil for "practice." This is obvious in law and medicine but, in a sense it is true of purely religious learning too. The study of sacred texts was regarded as desirable only if the person embodied the wisdom gained from them in his everyday life. Furthermore, the man made wise through study was expected to take a position of authority in his society. Thus the great scholars were expected to become leaders of the community, high civil servants, church dignitaries, judges. The humble scholar who lived for his learning was an exception, but even he was not honored merely for his scholarship. If he was not at the same time a saintly person whose private life was exemplary, he would not be honored. In his case, too, learning was a means to a practical end: the realization of the sacred way of life. Learning as such was not an end in itself. The conduct of learning was, therefore, an *amateur* rather than a professional activity; amateur teaching had higher prestige than the professional one. The transmission of religious and socially vital learning was not to be treated like a commodity that could be bought and sold on the market. This concept, of course, did not apply to the study of medicine or to secular studies like oratory or law, where payment was legitimate. But even in these areas, the brilliant practitioner-master who trained a few chosen disciple-apprentices, not the professional teacher, was the ideal.[2]

This pattern explains the rudimentary organization of learning in traditional societies. Where the teacher was first and foremost a practical person or a professional practitioner, he could become only marginally involved in a complex educational organization (as is the case even today in the clinical teach-

[2] Louis Ginzberg, *Students, Scholars and Saints* (New York: Meridian Books, 1958), pp. 25–58; Tielemann Grimm, *Erziehung und Politik im Konfuzianischen China der Ming Zeit (1368–1644)* (Mitteilungen d. deutschen Gesellschaft f. Natur u. Völkerunde Ostasiens, 1960), Vol. XXXV B; Jacob Katz, *Tradition and Crisis: Jewish Society at the End of the Middle Ages* (New York: The Free Press of Glencoe, 1961), pp. 183–198; Jacques Waardenburg, "Some Institutional Aspects of Muslim Education and Their Relation to Islam," *Numen* (April 1965), XII:96–138.

46

ing of medicine). This was a reasonable and efficient organization for legal-moral, religious, or even, up to a very developed point, medical and technological instruction. People who were capable of practical activity were not usually willing to give up their practice for teaching. Consequently, those preparing for a practical career or for a life of wisdom and saintliness preferred to become apprentices of first-class master-practitioners rather than to study with second-class masters who specialized in teaching. The exceptional creative teacher or thinker could assert himself anyway.

This state of affairs prevented any marked specialization and, in particular, specialization in theoretical studies. Specialization could not develop where a single master had to provide a comprehensive view of a whole field of learning and practice with only slight emphasis on his own preferred subject. Moreover, as long as the most respected teachers were practitioners, and the professional teachers were of low status, the practical applied approach prevailed over the theoretical one. The theorist, and therewith theory, like the person who was primarily a scholar, occupied a marginal position. Thus, natural science, mathematics, and even philosophy were marginal subjects. Even in ancient Greece and the Hellenistic world where philosophy attained a higher status and greater autonomy than elsewhere, its principal aim was still practical and moral.

Continuity of study had not been ensured under such conditions. Since the organization of teaching and study was quite informal, even famous seats of learning could decline very rapidly or even disappear without much struggle or conflict.

The European university was originally not different from the arrangements for higher learning of other traditional societies such as ancient India, China, or Islam. Students came from afar to Bologna, Paris, Montpellier, or Oxford in order to study law, theology, or medicine with famous masters, just as they would have gone to a famous master in India or Egypt.[3] In Europe, however, the conditions were different in one important respect. The towns where the famous teachers resided were autonomous corporations, and the foreign pupils were not under the protection of the king. The townspeople and scholars in Europe were often at odds. Violence was always close to the surface, and the history of the early universities up to the fourteenth century is full of accounts of unscholarly fights, murders, disorders and drunkenness. In addition to ineffective law enforcement, there was the problem of separation of church and state, with the church claiming responsibility and authority over all spiritual matters, including education, and denying the authority of secular

[3] The principal source of information for the medieval universities is Hastings Rashdall, *The Universities of Europe in the Middle Ages*, a new edition edited by F. M. Powicke and A. B. Emden (Oxford University Press, 1936), 3 vols. For the sociological aspects see F. M. Powicke, "Bologna, Paris, Oxford: Three Studia Generalia," pp. 149–179; "Some Problems in the History of the Medieval University," pp. 180–197; "The Medieval University in Church and Society," pp. 198–212; and "Oxford," pp. 213–229; in his *Ways of Medieval Life and Thought* (London: Oldhams Press, Ltd., 1950). See also Jacques Le Goff, *Les intellectuels au moyen age* (Paris: Seuil, 1957).

47

government over schools and scholars. This schism left the university scholars without firm regulation. To create order among the turbulent crowds of scholars and to regulate their relationship with the environing society, corporations were established. Students and scholars were formed into corporations authorized by the church and recognized by the secular ruler. The relationships of their corporation with that of the townspeople, with the local ecclesiastical officials, and with the king were carefully laid down and safeguarded by solemn oaths.[4]

The important result of this corporate device—which was not entirely unique to Europe but which attained a much greater importance there than elsewhere—was that advanced studies ceased to be conducted in isolated circles of masters and students. Masters and/or students came to form a collective body. The European student of the thirteenth century no longer went to study with a particular master but at a particular university.[5] A university consisted of several thousands of students (6,000 in Paris in 1300) and, at times, hundreds of masters living in an autonomous, intellectual community, relatively well endowed and privileged.[6] This intellectual community was much more independent of the pressure of society as a whole than single intellectuals serving the state or the church (or individual scholars working as teachers) could ever be. A man commissioned by several fathers or a religious community to teach a few students could rarely command much respect. But the same person in a university community of several thousands engaged entirely in the teaching of students could become, if successful, a very popular and admired man in that community. If the university was large, rich, powerful, and famous, his status in the larger society was very high too. Thus the specialized role of the university teacher emerged; it was a role enjoying high status. Furthermore, the dependence of status on intellectual and pedagogical accomplishment within the internal system of the university, rather than on practical services rendered to laymen, permitted specialization to a considerable extent in subjects of interest only to scholars.

Of course, in order to maintain its status, the university as a whole had to emphasize subjects that were important for society (such as law, theology, and medicine). But once they had become associated with a single institution, these subjects were placed in a new perspective.

The process occurred somewhat like this. Originally, the universities concentrated on one branch of study: law in Bologna, theology and philosophy in Paris and Oxford, medicine in Salerno, and law and medicine in Montpellier.[7] As long as the main institutional pattern of advanced learning was that of master and disciple, a famous doctor might have lived in one place and a great lawyer in another. Even if they had lived in the same place (e.g., in a capital city), they would not necessarily have had much to do with each other. How-

4 Rashdall, *op. cit.*, Vol. I, pp. 1–24, 43–73.
5 Powicke, *op. cit.*, pp. 149–179.
6 Rashdall, *op. cit.*, Vol. III, p. 355.
7 Rashdall, *op. cit.*, Vol. I, pp. 75–77 (Salerno); pp. 109–125 (Bologna); pp. 271–278 (Paris); Vol. III, pp. 1–5 (Oxford); Vol. II, pp. 115–139 (Montpellier).

the emergence of the scientific role

ever, once the universities became established, it became common for them to include all scholarly and professional subjects. It was easier to add another faculty to an existing corporation, and it was more worthwhile to do it in a place where there were undergraduates ready to enter the special faculties. These circumstances gave institutional support to the philosophical view of the existence of a coherent and comprehensive body of knowledge rendered meaningful by abstract theory. Thus, while law, medicine and, above all, theology continued to be the most important fields of learning, in the estimation of the layman within the university community itself, philosophy became the central subject. It was the basis of the intellectual culture common to all university-trained professionals.[8]

As a result, undergraduate study in the arts faculty—in effect almost exclusively scholastic philosophy—became in many ways the most important part of the university. This situation was true at least in Paris and Oxford, even though officially the arts faculty was only preparatory to higher studies in theology, law, and medicine. At the same time, philosophy became an increasingly specialized field in its own right and a highly respected one. In the twelfth and early thirteenth centuries it was still closely tied to theology, but soon schools of philosophy arose that became more difficult to accommodate within the traditional theology. The controversies concerning the variant interpretations of Aristotle are a good example. Their potential inconsistency with religious tenets was recognized early in the twelfth century and aroused theological opposition. The conflict was eventually resolved in the synthesis of St. Thomas Aquinas, but it reappeared again in a much more violent form when Averroist influences began to penetrate the universities. The doubt these influences cast on the immortality of the soul (and some other doctrines) led to harsh and extreme denunciations and interdictions of these theories by Bishop Tempier of Paris (1270 and 1277) and by Archbishop Kilvardby of Oxford (1277). Their reactions were the same as those the Averroist doctrine provoked in Islam and among Jews. But whereas theological reaction in these latter two instances attained its end of suppressing autonomous philosophy, this did not occur in Christian Europe. The universities were strong, and the philosophical faculty was the largest and most popular. As communities of professional experts, the universities could resist the authority of other professional specialists. Eventually Siger of Brabant and William of Ockham rationalized and legitimated the differentiation of philosophic from religious thought: philosophy had its own logic leading to necessary conclusions. The contradictions of religious revelation were not to be regarded as a refutation of philosophical arguments; they showed only the existence of a higher truth beyond human reason.[9]

This well-known story shows the importance of the differentiation and

[8] Le Goff, *op. cit.*, pp. 97–100, 108–133.

[9] A. C. Crombie, "The Significance of Medieval Discussion of Scientific Method to the Scientific Revolution," in Marshall Clagett (ed.), *Critical Problems in the History of Science* (Madison: University of Wisconsin Press, 1958), pp. 78–101; Le Goff, *op. cit.*, pp. 121–129; Guy Beaujouan, "Motives and Opportunities for Science in the Medieval Universities," in A. C. Crombie (ed.), *op. cit.*, pp. 219–236.

specialization which was taking place within the university. Philosophy became independent not because Ockham found the formula that made its independence somehow consistent with a "totalistic" religious outlook, but because philosophers had become a distinct and self-conscious group. That group had become large and respected enough to defend itself, which made it necessary to find a compromise formula. And they were a large and respected group because the universities were important and powerful centers of intellectual activity whose internal scale of values could not be easily disregarded.

The importance of this development lies precisely in a circumstance that is usually least appreciated: this development was originally a purely intellectual revolution. Unlike the rise of the philosophical schools in Greece and China, this philosophy did not compete with, nor did it ever replace, traditional religious doctrine. It only established a limited measure of professional freedom and equality for a new group of university intellectuals interested in a study that was neither practical nor religiously sanctioned. From the point of view of the fight for freedom of thought, as well as the intrinsic value of the philosophy thus emancipated, the achievement may appear petty. But from the point of view of the development of science and, in fact, from that of the development of institutional conditions for freedom of thought, it is exactly this apparent "pettiness" which is important. Complete victory would have led to the emergence of a new totalistic world view, competing for worldly power and influence with the religious one. The establishment, as well as the eventual overthrowing of such a philosophy, would have necessitated bloody revolutions. Since this did not happen, a first step was taken toward the separation of intellectual from political-religious revolutions. This separation was a necessary precondition of the emergence of science as an independent intellectual field.

The Peripherality of Science
in the Medieval University

The intellectual division of labor arising from having different kinds of studies within one corporate organization also stimulated the further internal differentiation that gave the natural sciences their place at the universities. They were not a required part of the curriculum, and any academic degree could be acquired without knowledge of them. But inevitably among so many masters and scholars studying a considerable variety of subjects, there were some who were interested in scientific problems. Logicians took up mathematical and physical problems, and physicians considered a variety of biological problems. Informal groups emerged and were given facilities to pursue these studies outside the regular curriculum or during vacations.[10] Even though these activities were not institutionalized, the mere size and internal differentiation of the universities permitted enough interested persons to find each other. In such a large academic "market," there was enough "demand" to maintain even

[10] Beaujouan, *op. cit.*, pp. 220–224, 233–236.

50

the emergence of the scientific role

a marginal intellectual field. In a small circle, by contrast, the probability of finding anyone interested in it would have been less and the stimulus to curiosity and persistent interest would have been correspondingly small.

Decentralization played its part in all these processes. The corporate autonomy of the university in any single place would not have been sufficient to withstand the onslaught of church authorities against the philosophers. Nor could, or did, scientific activity survive the disruptions caused by war, plague, and political strife in any single place. But during the thirteenth century when the opportunity to move diminished because of increased royal power (especially in Paris), it was still possible for individuals to go to a different country, as did the English and the Germans, and even some other scholars (e.g., Marsilius of Padua and John of Jandun) who left Paris for English or German universities.[11] The decline of philosophy in France and England made Italy the new center of its study in the fifteenth and sixteenth centuries. The intellectual differentiation continued, and the fifteenth century Italian humanists or the sixteenth century neo-Aristotelians were completely secularized and specialized philosophers.[12]

Because geometry and dynamics were mainly up to the sixteenth century cultivated by philosophers, the fate of these studies was bound to that of philosophical studies in general. The tradition of medieval natural science was started at Oxford by masters of Merton College. From there it spread to Paris, which had the closest intellectual commerce with Oxford. When the tradition declined in both places during the fourteenth century, as did philosophy, the center shifted to Italy, mainly to Padua, and to the new German, Dutch, and other universities.[13] Thus, when special university chairs were established, in the fourteenth and fifteenth centuries this tradition, influenced probably by internal developments within the medical faculty, also led to the establishment of professorships in mathematics, astronomy, and a variety of subjects, such as natural philosophy, Aristotelian physics, and so forth, first in Italy and later everywhere in Europe. These scientific chairs were of subordinate importance; it was an advancement for their incumbents if they could be appointed as professors of philosophy, or even better, of theology, law, or medicine. In any case, it was necessary to have a degree in these latter subjects in order to be appointed to a chair. Nevertheless, by this time the natural sciences were more or less regularly taught—on however modest a level—by professors who were paid for teaching them.[14] (See Table 4-1.)

[11] Richard Scholz (ed.), *Marsilius von Padua, Defensor Pacis* (Hannover: Hahnsche Buchhandlung, 1932), pp. lvii–lxi.
[12] John H. Randall, tr., *The School of Padua and the Emergence of Modern Science* (Padua: Edifice Antenore, 1961), pp. 21–26.
[13] Le Goff, *op. cit.*, pp. 156–162, 167–176.
[14] Albano Sorbelli, *Storia della Universita di Bologna* (Bologna: Nicola Zanichelli, 1940), Vol. I, Il Medievo (Secc. XI–XV), pp. 122–126, 252–254; A. R. Hall, "The Scholar and the Craftsman in the Scientific Revolution," in Clagett (ed.), *op. cit.*, pp. 3–23; H. Helbig, *Universitaet Leipzig* (Frankfurt a. M.: Weidlich, 1961).

Table 4–1

Number of Salaried Chairs at Selected Universities by Fields, 1400–1700

	1400	1450	1500	1550	1600	1650	1700
Bologna							
Science	3 [a]	—	2 [b]	2	2	2	2
Medicine	11	2	3	3	5	5	3
Other	33	9	15	16	20	22	23
Paris (Sorbonne and Collège de France)							
Science	—	—	—	2 [c]	2	2	2
Medicine	—	—	—	2	3	3	3
Other	—	—	—	8	12	18	20
Oxford							
Science	—	—	—	—	—	3 [d]	3
Medicine	—	—	—	1	1	2	2
Other	—	—	—	15	15	20	20
Leipzig							
Science	—	—	—	—	2 [e]	2	2
Medicine	—	2	2	3	4	4	6
Other	—	—	—	—	17	17	23

[a] Astrology; natural philosophy; physics.
[b] Arithmetic and geometry; astronomy.
[c] Mathematics.
[d] Natural philosophy; geometry; astronomy.
[e] Arithmetic and astrology; physics and natural philosophy.

SOURCE: Sorbelli and L. Simeoni, *Storia della Università di Bologna* (Bologna: Università di Bologna, 1944); A. Lefranc, *Histoire du Collège de France* (Paris: Hachette, 1893); J. Bonnerot, *La Sorbonne* (Paris: Presses Universitaires, 1927); C. E. Mallet, *A History of the University of Oxford* (New York: Longmans, 1924); H. Helbig, *Universität Leipzig* (Frankfurt a. M.: Weidlich, 1961).

This process of differentiation, however, was stabilized some time in the sixteenth century (varying from country to country). By then, universities introduced the teaching of classics, basing their arts curriculum on the new set of specialized disciplines advocated by the humanists. There were in addition a number of professors of mathematics, astronomy, and natural history. Finally, there was a measure of specialization in the medical faculty; astronomy was also a specialized field of study, and the idea of basic medical sciences was accepted. This set of subjects remained almost unchanged essentially until the end of the eighteenth century. The only further differentiation of importance to science occurred in the medical faculties where chemistry became a relatively important and definitely specialized field during the eighteenth century. The status of mathematics and the natural science disciplines remained low. The natural sciences did not have anything approaching the status of the humanistic

the emergence of the scientific role

subjects, not to speak of the status given subjects of the professional faculties.[15] This circumstance greatly limited the autonomy of the scientist's self-image. As long as a scholar could obtain a university position in mathematics only if, in addition to mathematical knowledge, he possessed a degree in medicine (e.g., Cardan), theology (e.g., Luca Pacioli), or law, and as long as advancement and fame depended on his being not only a competent mathematician but a good classical scholar, there was little incentive to concentrate his energy on scientific subjects. One or another highly motivated scholar might decide to devote his best talents to science. But there was no institutional basis to ensure that his successor would do likewise.

The situation was not much different in medicine, although the status of that faculty was higher than that of the faculty of arts. Anatomy was considered an integral part of the subject, and chemistry was integral to the apothecary's art, which fell within the jurisdiction of the medical faculty. Here too, however, the important phase was the practice and theory of medicine; the scientific aspect was less important. An anatomist who was not a physician— there were some artists of this kind—was not seriously regarded. Chemistry, though an undoubtedly more developed science than any branch of medicine or biology by the second half of the eighteenth century, was lower in prestige and importance than those two subjects. Its status corresponded to the lower art of the apothecary. As a result, there was little continuity of science in the medical faculties. Continuity was largely a function of the accidental interest of individuals. However, the successor of a great anatomist-physiologist could be a practitioner with no scientific interests. Moreover, the level of chemistry teaching varied a great deal, since the technical aspects of the subjects were not generally appreciated as being of academic importance. (This is why, even as late as the eighteenth century, the scientific excellence that occurred in this field in the German, Dutch, Swedish, Scottish and Swiss universities was neither general in all fields nor of long endurance.) [16]

These limitations on the growth of science in the universities have been explained in a variety of ways. The decline of Oxford and Paris in the second half of the fourteenth century has been attributed to the Black Death and the Hundred Years' War. At different times during the sixteenth and the seventeenth centuries, German and English universities were subject to religious purges. However, these were local and temporary phenomena which had ceased to be of importance in these countries by the second half of the seventeenth century. In other countries, such as France, religious purges had ended as early as the fifteenth century. Nonetheless, the fact remains that from the fifteenth

---

[15] Nicholas Hans, *New Trends in Education in the Eighteenth Century* (London: Routledge & Kegan Paul, Ltd., 1951), pp. 47–54; Stephen d'Irsay, *Histoire des Universités françaises et étrangères* (Paris: Auguste Picard, 1933–1935), Vol. II, pp. 108–118.

[16] Leonardo Olschki, *Geschichte der neusprachlichen wissenschaftlichen Literatur* (Leipzig, Firenze, Roma, Geneve: Leo Olschki, 1919), pp. 414–451; (Halle an der Saale: Max Niemeyer, 1927), Vol. III, pp. 95–96, 105–107; Joseph Ben-David, "Scientific Growth: A Sociological View," *Minerva* (Summer 1964), II:464.

century on in Europe as a whole (including Italy where the universities flourished), the major contributions to science (except in medicine) were made outside the universities.

The reason for this halt in the development of science in the European universities must, therefore, be sought in something that was common to the whole system. This common factor cannot be some general decline and corruption of university standards, since in some fields, notably law, the universities continued to be centers of creativity. Even in science there were certain exceptions. Medicine and its basic sciences, for example, fared better in the universities than physics and mathematics.

The explanation seems to be that the decline of the role of the universities in the development of science had been due to general limitations imposed by the place of the university in society. The emergence of new functions within an organization requires the integration of the new parts. In principle this integration can be accomplished either by creating a hierarchy between the functions that subordinate one to the other according to some external criterion, or by coordinating the new and old functions. Whether one or the other alternative is chosen depends on the social uses of the activities concerned.

Because the uses of university study were to supply lawyers, civil servants, clergymen and doctors, the organizational decision had to be in favor of the subordination of science to general philosophy, classics, and professional studies. Philosophy and classics could be raised to a status of near equality with the professional studies, since they could be regarded as necessary preparation for the latter. If the universities made philosophy and classics the central part of their studies, this emphasis only meant a preference to master the methods and the tools of professional study over its substance. A person whose Latin was good and who knew the laws of logic could find his way in the legal or theological compendia and was able to learn the rules of his profession without undue difficulty. Even for those studying medicine it was considered more important that they be able to read Galen than to study anatomy and physiology. Certainly it was more of an intellectual feat to master Latin and Greek and scholastic or classic philosophy than to learn the few things known at that time about basic medical science.

When the function of the university up to the end of the eighteenth century is recognized, the reason that it could not offer more than a subsidiary place for the sciences can be understood. It was not possible to make a good case for the extension of science at the universities either from the point of view of general or professional education until the end of the eighteenth century. The task of teaching at a university, therefore, was not conducive to raising the aspiration of the scientists or inspiring a new world view in which natural science was the paradigm of all philosophical knowledge. Having been established as a distinct and differentiated philosophical strain within the university, the social basis for the complete autonomy of science had to be found

54

elsewhere. At any rate, more favorable conditions for the attainment of scientific autonomy emerged outside the universities.

Artists and Scientists in Italy: The Rudimentary Formation of the Scientific Role

The first signs of a change in the evaluation of science appeared in the circles of artists and engineers in fifteenth-century Italy. Until then, artists were considered mere artisans, but, as a result of the general conditions that made possible a modicum of autonomy for various urban groups, their fortunes were improving in the fifteenth century.[17]

In addition to the new interest in art, this improvement in status was perhaps even more closely related to the fact that the role of artist often overlapped, in the same person, with the roles of architect, fortification engineer, and ballistic expert. In fifteenth-century Italy, the artist received an all-round training. As a youth apprenticed to the workshop of a master, he tried his hand at painting, sculpture, architecture, and goldsmithery before he specialized. If he was outstanding, he entered the service of a city or of a secular or an ecclesiastical prince to be responsible for public works in art, architecture, and engineering. Verrocchio, Mantegna, Leonardo da Vinci and Fra Giocondo were among these versatile artist-technicians. They received a superior kind of technical training, which was comprehensive and eminently practical. But the artists had little formal education. Usually, they did not know Latin, and whatever book knowledge they had could have come only from the popular compendia in the vernacular which attempted to digest all available knowledge in an uncritical fashion. There had already been communication between scholars and architects before the fifteenth century, the latter consulting the scholars about classical technological manuscripts. But starting from the first half of the fifteenth century with the school of Filippo Brunelleschi—which included Luca della Robbia, Donatello, and Ghiberti—these communications between scholars and architects became more continuous and conventional. The school of Brunelleschi included Leone Battista Alberti, a rich and learned scholar who became an architectural theorist and consultant to the group.[18]

The connections between the artists and the university-trained scholars were based partly on common technical interests. The artists and architects were interested in problems of perspective, the engineers in statics and dynamics. They could both benefit from the scholars who knew the available classical literature and could express in articulate principles what the artists could not. At the same time the scholars benefited from their connections with

[17] Olschki, *op. cit.*, Vol. I, pp. 21–44 and Vol. III, pp. 414–451.
[18] Olschki, *op. cit.*, Vol. I, pp. 33–36, 46–88; Giorgio de Santillana, "The Role of Art in the Scientific Revolution," Clagett (ed.), *op. cit.*, pp. 33–65.

the artists whose practical experience helped to make the content of the ancient texts meaningful. Greek geometry and science became more intelligible when studied as part of design, construction, or ballistics rather than as pure book learning. The interest of painters in anatomy and botany provided a powerful tool for the anatomists and naturalists.

The artists had certain status problems in common with scientists. Both artists and technologists had hitherto held relatively low positions in society. The only practical way to assert the status of their calling and to prove the spiritual value of what had been traditionally considered as a lowly manual art was through evidence of the connection between their work and a recognized scholarly pursuit. However, they were not very interested in the acquisition of classical languages and had no sense for philosophical speculation. The only scholars with whom they had a common interest were those cultivating the sciences.[19]

This identity gave rise to a new image of the scientifically inclined scholar. In the university community on which his status had until then depended entirely, his interests had been considered of mere peripheral importance. If he wanted to obtain recognition, he had to prove his worth in the more central fields of scholarship. Beginning in the fifteenth century, however, there was an upcoming profession, that of the artists, for whom philosophy was primarily science. Viewing themselves through the eyes of these new clients or public who appreciated what they had to offer, the scientist-scholars gained self-confidence. There was a basis for viewing science and mathematics as the center of a new philosophy still to be created.

As a result a continuous interchange between artists and scientists developed that lasted throughout the fifteenth century.[20] Alberti's and Brunelleschi's attempts to create a science of art and architecture were followed by many artists such as Ghiberti, Antonio Averlino Filarete, Francesco di Giorgio Martini, Piero della Francesca, Leonardo da Vinci, and Dürer. Even more important than these men to the development of science were the trained scholars having connections with artists. Paolo dal Pozzo Toscanelli, Luca Pacioli, Cardan, Bernardino Balbi, Tartaglia, Peurbach, and Regiomontanus all had associations with

[19] Paolo Rossi, *I filosofi e le macchine (1400–1700)* (Milano: Feltrinelli, 1962), pp. 11–12, 21–31, 40–42.

[20] In order to avoid cumbersome expressions I shall use "scientist" rather than "scholars specializing in or interested in science." It is, however, important to keep in mind that scientists as a social and intellectual category distinct from scholars did not exist before the seventeenth century. The people we are dealing with were scholars who had an increasingly articulated feeling that their scientific interests did not fit into existing intellectual schemes; they became very gradually aware of the possibility of viewing themselves as something different from scholars. The fact that they held distinct chairs at the universities should not be taken as a sufficient criterion of the existence of a new profession or even of the existence of a distinct identity of "scientists." The chairs were unimportant ones and their incumbents were regarded as experts in a subspecialty of philosophy or medicine, not as specialists in subjects of autonomous dignity.

the emergence of the scientific role

artists or at least used their work. The connection was even closer in anatomy and botany.[21]

The importance of the experience of artists and artisans to the development of scientific ideas has often been debated. Some historians attach very great importance to them and others point out that the decisive discoveries from Copernicus to Newton were all made by trained scholars and were derived from the existing intellectual traditions of the Middle Ages (e.g., the impetus theory) and rediscovered classic works.[22] Whatever the merits of these arguments, there is no doubt that these associations were important in making the social image of the scientists distinct from other scholars and in conferring new dignity upon scientific activity. In the circles of artists and engineers, which toward the end of the fifteenth century centered in the courts of Prince Frederico at Urbino and of Ludovico Sforza in Milan (at this latter court, Leonardo da Vinci was the central figure), there developed a conception of the individual genius whose knowledge did not stem from books but from his personal intuition on the one hand and his contact with nature on the other. Frederico of Urbino was one of the great Renaissance princes with a great library and wide-ranging interests. And at the court of Ludovico Sforza in Milan there was a circle (at times referred to as an academy) which included Leonardo da Vinci; Gometio, a theologian; Domenico Ponzone, a preacher and head of a cloister; Ambrogia da Rosate, astrologer and court physician; Alvise Mailiani, university professor in Pavia, mathematician, theologian and poet; Gabriele Priovano, rector of Pavia University; Niccolo Cusano; Andrea Novarese; Galeazzo di Sanseverino, *condottiere* and military engineer; and Luca Pacioli, mathematician. (The last then left for Florence and stayed there with Leonardo da Vinci after the fall of Ludovico Sforza).[23] Among these groups of recognized and respected intellectuals at two of the important courts of the time, artists, engineers, and scientists were accepted as the equals of established scholars and theologians. Acceptance in the constitution of such a circle implied recognition of the dignity of scientific activity; it acknowledged the inherent value of devoting oneself to scientific work.

These groups were of decisive importance in the events that led to the dramatic aspiration of Galileo a century later to obtain the recognition that scientific work constituted the central element in the role of the philosopher as he ought to be. Galileo felt that this role of the new philosopher was equal in definition and dignity to the other well-established intellectual roles, such as the legal or theological expert, the physician, or the humanistic scholar. But the aspiration, at least in Italy, ended in failure.

[21] Olschki, *op. cit.*, Vol. I, pp. 109–127, 151, 159–161, 199–200, 414–451.
[22] A. C. Keller, "Zilsel, the Artisans and the Idea of Progress in the Renaissance," *Journal of the History of Ideas* (1950), XI:235–240; Alexander Koyré, "Galileo and Plato," *ibid.* (1943), IV:400–428.
[23] Olschki, *op. cit.*, Vol. I, pp. 156–161, 239–251.

57

Although we cannot reconstruct the complete chain of events, there is enough information available to sketch their outline. The hybrid identity of artist-scientist was a transient phenomenon. By the beginning of the sixteenth century, the painters and architects had learned everything that could be useful to them in geometry and optics, which was not a great deal. With Michelangelo a reaction set in against the confusion of art with science.[24] Scientists could still benefit from association with engineers (as well as with artisans such as lens grinders and instrument makers). But these were very specific and technical skills and could not provide anything like the revelation that the fifteenth century students of Euclid and Archimedes must have experienced when they discovered that, in the work of artists and engineers, geometry and mechanics obtained a new dimension and vitality never attained in the learned discussions of their academic colleagues. Nor could these new skills match the discoveries of anatomists who first taught the artists how to draw a human body as actually perceived. By the middle of the sixteenth century, however, the relationship between science and art reverted to the earlier pattern of two endeavors running widely separate courses and having few meaningful encounters. To the extent that there was continued association, there was nothing in it to introduce a new element in the situation. The association that had been a revelation in the previous century developed into a mere routine.

Meanwhile, starting from the 1530s in the northern countries of Europe, there emerged a growing trend to extol the virtues of arts and crafts and the knowledge of nature. This trend began in the writings of Ludovico Vives, Erasmus, Montaigne, and Rabelais and can be traced through Palissy to Bacon's new philosophy.[25] This intellectual trend ran hand in hand with the continued growth in the social importance of new classes whose outlook was sympathetic neither with the scholastic nor humanistic intellectual establishments. In Italy, on the other hand, artists and technologists had been unable to break away from the domination of the guilds in spite of attempts in this direction. Scientists as well as the small number of very eminent artists were now moving in a quite different, upper class, humanistic environment—the environment of the academies.[26] This environment embraced those merchants who were being absorbed into the nobility in Italy. This development was due partly to the nature of Italian city-state democracy and partly to the fact that, unlike the position in Northern Europe, there were no Protestants with great influence in these important classes to foster an opposing intellectual outlook potentially congenial to science. Thus at a time when the Northern European class structure became increasingly fluid and a mobile middle class was increasing in size, in awareness

24 *Ibid.*, Vol. I, pp. 255–259.
25 Rossi, *op. cit.*, pp. 11–12.
26 Nikolaus Pevsner, *Academics of Art: Past and Present* (Cambridge: Cambridge University Press, 1940), pp. 50–66.

58

of itself, and in self-sufficiency, the Italian class structure recrystallized into something approximating its earlier form.[27]

The Reconquest of Science
by the Nonscientific Culture
in Italy

This description of the Italian class system as an increasingly rigid one and that of Northern Europe as an increasingly fluid one should be qualified. So should the parallel description of Italian science as approaching stagnation and Northern European science as developing at a growing rate. Looking backward from the vantage point of the second half of the seventeenth century, the statement appears correct, but viewing the situation from the perspective of the sixteenth and early seventeenth century, it seems misleading. The merging of the merchants into the Italian nobility may be considered as a sign of open-mindedness toward commercial occupations that was not known in the majority of European countries before the nineteenth century. The participation of guilds in the governments of cities ensured a wider extension of civic rights than anywhere else, and the interest in science, as well as in all other branches of learning and art, was more widespread in Italy than in any other country. In what sense is it then justified to view the turning of the scientists to the upper class social framework of the academies as a foreshadowing of decline? Wouldn't it be more correct to interpret it as the first appropriation of a new scientific culture by part of one of the ruling classes in Europe? (This was a move to be followed later by other countries.)

For our purpose, the main point is that in other parts of Europe the cause of science was taken up by a class of persons who stood to gain from changes in the social order. In Italy by contrast, science became, by the sixteenth century, the concern of a minority within a class which had attained what it wanted and which was interested in social stability.

After a period among the artists, the Italian scientists began to feel strong enough to seek the recognition of the official intellectual community. This recognition was eventually denied them by the ruling circles of the church and the state as well as the intellectual establishment. In this process of overture and rejection so essential to an understanding of the stagnation of Italian science, the Italian academies played a great part.

The academies developed from the intellectual circles that had grown up about 1440 around famous humanists like Rinuccini and Ficino in Florence and Pomponio Leto and Cardinal Bessarion in Rome. Originally the circles were wholly informal groups for the discussion of the revived Platonic philosophy and the whole range of humanistic learning, science, vernacular literature, and

[27] C. M. Cipolla, "The Italian and Iberian Peninsulas," in *The Cambridge Economic History of Europe*, Vol. III (New York: Cambridge University Press, 1966), pp. 397–430.

arts. In the beginning, they were not specialized, and their typical form was either a master and a circle of disciples, or a group of intellectuals enjoying the patronage of a great magnate or a prince.

The term "academy" was programmatic: it expressed the intention of the founders of the Platonic academy in Florence in 1454 to compete with the old Aristotelian tradition of the universities, which they regarded as uncongenial. Those who revolted were not intellectual outsiders but university-trained philosophers (some of them with scientific interests), classicists, jurists, and physicians, with a preponderance of interest in subjects pursued in the arts faculties. They had gained entrance into the church and the courts, and many of them were wealthy and powerful people themselves.[28]

The emergence of effective rulers in the Italian cities who could manipulate and control the guilds, and of an upper class of rich bankers and merchants, made it possible for intellectuals dissatisfied with the atmosphere of the university (which, in the beginning, resisted the new learning), to create their own groups in rivalry to the official corporations. This formation of intellectual groups outside the universities was only a continuation of what had started in the universities, that is the active search for the classical heritage and the development of various lines of specialization within the existing tradition. The Platonic revolution of the fifteenth century was not very different from the Averroist revolution of the thirteenth century. Both were brought about by professional intellectuals for primarily intellectual interests. The differences lay in their institutional form, focus, and their relations to authority. The thirteenth century revolution could take place only in the universities. But by the fifteenth century, to avoid a head-on collision with established ways of doing things within the universities, it was possible to withdraw into a circle of colleagues, disciples, and patrons. Masters and disciples no longer needed the protection of a guild of their own, and thus the interference of the guild in their activities lost its rationale. Nor were they dependent on any ecclesiastical benefices or clerical, or quasiclerical, privileges. It was possible to obtain more or equally effective protection and support from princes, great noblemen, or even municipalities. It had also become easier to enjoy intellectual conviviality with adults and equals without being in a clerical order. To some extent, of course, this was no more than the appearance in the rich northern Italian cities of a circle of court intellectuals as had existed earlier in the Orient and in Spain. The differences lay in the autonomous, corporate character of the European academy and in its intrinsically intellectual orientation. It consisted not so much of individual scholars seeking the protection and patronage of a king as of groups of equals seeking an appropriate forum for intellectual intercourse (though still in need of princely patronage).

The first century of the academies may be interpreted as essentially an

[28] Martha Ornstein, *The Role of Scientific Societies in the Seventeenth Century* (Chicago: The University of Chicago Press, 1928), pp. 73–90; Pevsner, *op. cit.*, pp. 1–24; D'Irsay, *op. cit.*, Vol. I, p. 226 ff. and Vol. I', pp. 45–128.

the emergence of the scientific role

attempt of persons, many of whom otherwise might have been forced to work through established university faculties of arts, to create for themselves an intellectually more congenial institution than the universities provided. They did so by making use of the new resources of wealth and protection in such centers as Florence, Rome, Naples and, later, Paris and London.[29] The numbers of persons with such interests were increasing; many of them did not have to gain their livelihood from teaching. They sought to enrich their understanding by coming together to discuss matters of common interest.

Until the middle of the sixteenth century, however, the academies took no more interest in science than the universities had. Those circles which did take an interest in science (the circles of artists and scientists flourishing at Urbino and Milan at the turn of the fifteenth century) were not generally considered academies. Whatever the actual status of the artists, the circles that they formed were unable to lay claim to the prestige-bearing designation of "academy."

During the first century of their existence, the academies tended to embrace almost the entire range of intellectual activities. After the middle of the sixteenth century, general academies with a wide diversity of purposes were founded with less frequency. Specialized academies started to spring up, of which half or more were literary academies with the rest being divided among theatrical, legal, medical, theological, scientific, and artistic purposes (see Table 4–2).

There was also a significant change in social structure. Instead of being relatively informal groups, the academies became increasingly formal institutions, conferring upon their members publicly recognized honors. Among other indications, this was manifested in the composition of the membership in which noble amateurs tended to outnumber professional intellectuals. This tendency toward formalization occurred in the literary and multipurpose academies in the middle and end of the sixteenth century, but formalization in the scientific academies occurred only at the end of the seventeenth and in the eighteenth centuries (see Table 4–3).

These changes indicate the success of the fifteenth-century movement of academy foundation. The humanistic studies that were promoted by the academies were taken over by the universities; where the faculty resisted, new institutions were established, like the Collège des Lecteurs Royaux (later Collège de France), in Paris.[30] The foundation of multipurpose academies ceased when the academy lost its function as a counterfaculty in which intellectuals sought refuge when they were unable to find satisfaction in the narrow scholastic atmosphere of the universities. The academies continued to thrive where the universities failed to step in, namely, in the study and support of vernacular language and literature. There was increasing national pride in these studies, and rulers backed the trend for political reasons. At the same time, these subjects were

---

[29] The patrons of the more ambitious and important groups were Cosimo and Lorenzo di Medici, Alfonso I of Aragon, Ludovico Sforza, and others; some of the intellectuals were also powerful figures, for example, Cardinal Bessarion and Count Cesi (Pevsner, *loc. cit.*).

[30] D'Irsay, *op. cit.*, Vol. I, pp. 270–274.

61

the emergence of the scientific role

Table 4–2

Percentages of Foundations of Various Types
of Italian Academies, 1400–1799

	Literature	Science	Medicine	Law	Divinity	Theater	Multi-purpose [a]	Total	No.
1400–1424	—	—	—	—	—	—	100.0	100.0	1
1425–1449	—	—	—	—	—	33.3	66.7	100.0	3
1450–1474	—	—	—	—	—	50.0	50.0	100.0	2
1475–1499	16.7	—	—	—	—	16.7	67.7	100.0	6
1500–1524	29.4	—	—	—	—	35.3	35.3	100.0	17
1525–1549	55.6	3.7	—	—	—	7.4	33.3	100.0	27
1550–1574	58.8	7.6	1.5	2.9	—	13.2	16.2	100.0	68
1575–1599	61.2	2.0	4.1	2.0	2.0	14.3	14.3	100.0	49
1600–1624	59.6	3.4	1.1	0.0	3.4	16.9	15.7	100.0	89
1625–1649	53.6	2.4	2.4	2.4	4.8	26.8	7.3	100.0	41
1650–1674	60.3	9.5	0.0	1.4	2.7	16.2	9.5	100.0	74
1675–1699	53.7	9.3	1.9	3.7	9.3	11.1	11.1	100.0	54
1700–1724	67.7	3.1	0.0	1.5	6.2	9.2	12.3	100.0	65
1725–1749	51.0	2.0	2.0	0.0	27.5	11.8	5.9	100.0	51
1750–1775	59.3	5.1	3.4	1.7	15.3	6.8	8.5	100.0	59
1775–1799	51.0	3.9	2.0	5.9	7.8	15.7	13.7	100.0	51
Total	56.3	4.7	1.5	1.8	6.7	14.5	14.3	100.0	657

[a] All academies that pursued any combination of more than one interest. About half of this category also pursued some scientific interests.

SOURCE: M. Maylender, *Storia delle Accademie d'Italia* (Bologna: L. Capelli, 1926–30). An approximately 50 percent sample, consisting of letters A–C, R–Z: Vol. I; Vol. II, pp. 1–150; and Vol. V; excluding all double entries, schools called "academies," and unclassifiable entries.

not yet considered serious enough to be important in the rigorous education of mind and taste that the universities undertook, nor could they replace Latin as the language of theological, legal, medical, or philosophical learning. Thus instead of being made important parts of the university curriculum, vernacular language and literature were relegated to the academics. Writers were honored with titles whose value was attested by the interest of the nobility in obtaining them.

The academies provided a flexible framework for the expression of the cultural interests of different groups of intellectuals where those interests could not be fulfilled by existing institutions. The founding of such institutions in Italy as a means of coping with newly emerged interests would seem to indicate the relative openness of Italian social structure as compared with the rest of Europe where academies were created only in imitation of the Italian models.

As a matter of fact, however, what appears as openness is better interpreted as evidence of rigidity. A price had to be paid for the relatively easy absorption

the emergence of the scientific role

Table 4–3

Numbers of Multipurpose and Scientific Academies
in Italy, 1430–1799, Classified by Structure

	MULTIPURPOSE [a]			SCIENCE		
	Total New Foundations	*Informal*	*Formal*	*Total New Foundations*	*Informal*	*Formal*
1430–1479	6	3	3	—	—	—
1480–1529	6	4	1	—	—	—
1530–1579	23	6	15	9	6	3
1580–1629	14	5	7	6	2	2
1630–1679	6	1	5	14	10	3
1680–1729	10	2	8	17	5	11
1730–1779	13	1	11	8	3	4
1780–1799	12	2	10	5	1	3
Total	90	24	60	59	27	26

[a] Includes only those of the multipurpose academies which also included science among their various fields of interest.

SOURCE: Maylender, *op. cit.* Vols. I–V, 100 percent, sample. "Informal" academies include: a patron with surrounding circle of intellectuals; a famous intellectual and circle of disciples; or a group of intellectuals gathering for informal discussions. "Formal" academies are of the following types: an organized professional association; a group of noblemen holding regular meetings at which the intellectuals whom they patronize give lectures or demonstrations; or an honorific group of nobles and intellectuals. All of the latter three types tended to have elaborate offices, rules, a seal, a motto, academic names for the members, etc. Discrepancies between total foundations and sums of the two structural types are due to a small number of unclassifiable academies (six in the multipurpose academies and six of the science academies).

of the leading merchants into the nobility and for the relative ease with which the new cultural pursuits were accommodated within academies and the academies within the official hierarchy of cultural institutions. The price was the assumption of the habits of thought, attitudes, and style of the upper classes to the point where the spirit of innovation eventually expired.

One of the results was the abandonment of the practical concerns of science. Whereas propaganda in England and France for the official recognition of science was based on its potential usefulness to technology and production, in Italy its claims were justified by arguments from Platonic philosophy or neo-Platonic mysticism. The cause of science in Northern Europe was supported not only by certain, usually upper class, intellectual circles actually cultivating it, but also by a considerable number of merchants, artisans, and seafarers. In Italy science was espoused only by an upper class intellectual clique that was trying to displace the official university philosophers and modernize the intellectual outlook of the Catholic Church.[31]

[31] "Upper class" is used here in a somewhat loose way. British and Dutch merchants and seafarers were often of upper class origin, while some of the scientists were not. But as a status category merchants were not upper class in the West, while official academies were upper class institutions. In Italy big merchants were upper class too.

Copernican astronomy was the issue around which conflict crystallized in Italy between the opposition and the official intellectual establishment. It had obvious philosophical implications that were useful for the opposition movement, but in the end it embroiled the movement with the Church. The oppositional as well as the conspirational and esoteric nature of the movement is attested by the names of the academies in the sixteenth century: Incogniti (Naples, 1546–1548), Segreti (Naples, 1560), (Vincenza, 1570), (Siena, 1580); Animosi (Bologna, 1562), (Padua, 1573); Affidati (Bologna, 1548). The reappearance of the same names in different places probably shows the existence of links between certain groups in several localities. None of these names reappear in the next centuries when the movement came out into the open and renounced its oppositional orientation.

The first group of significance was perhaps the short-lived Affidati in Padua. Founded in 1573 by the Abbot Ascanio Martinengo, it included professors of the university, high clergy, noblemen, and internationally known scholars. It did not last long, but some of its members went to Rome where some twenty years later they appeared as members of one of the most famous academies, the Accademia dei Lincei. The latter was founded in 1603 by the eighteen-year-old *Marchese* Cesi who was joined in 1610 by the Naples physicist Giambattista della Porta, whose academy in Naples had been suppressed by the Roman Curia. In 1611 Galileo, who had previously resigned his unsatisfactory chair at Padua, also joined the Accademia dei Lincei. This circle can be regarded as the first that made an open and relatively comprehensive attempt to create a scientific institution claiming equal status to other institutions of learning. The Lincei attempted to organize instruction in natural sciences, philosophy, and jurisprudence, and it published books on science, including two by Galileo.[32]

It is very doubtful whether the dramatic events of the condemnation of Galileo were in themselves of far-reaching significance. The indignation aroused by the high-handed actions of the church hierarchy probably increased the popularity of science. There is indeed no sign of a cessation of scientific activity following the condemnation of Galileo. It is true that the activities of the Lincei were curtailed after the first attempt of the Curia to suppress Galileo. But some of its members and the disciples of Galileo continued to be active throughout the first half of the century, and they participated in the foundation of the other famous Italian academy of the seventeenth century, the Cimento (1657–1667). After its patron, Prince Leopold de Medici, was elected a cardinal, its members, due to personal animosities, were unable to carry on with their work.[33]

The picture, therefore, is not one of a movement that gained wider and wider support and then was violently suppressed, but rather of an episode well contained within an established intellectual fraternity which was decaying. By

[32] Ornstein, *op. cit.*, pp. 74–76.
[33] A. Rupert Hall, *From Galileo to Newton* (London: Collins, 1963), p. 135.

the emergence of the scientific role

the end of the seventeenth century, the scientific academies had become unimportant replicas of the literary academies, consisting of local amateurs and notables. They were of no importance in international science. In the field of medicine, Italy remained a center during the late seventeenth century thanks to the excellence of some of its university faculties. But in other sciences the center shifted to England and France. The same fate that had befallen English and French universities at the turn of the fourteenth century when they lost their leadership to Italy now befell the Italian academies. Science and scientists remained dependent on the narrow circles within the upper classes that ruled both the country and the church and were interested in learning. These were the circles which had to be convinced that natural science was important and worthwhile enough for them to give to it the full blessing of public recognition and freedom of communication notwithstanding the weighty doctrinal difficulties that might arise from such recognition. The circles were not convinced however. Indeed, it could not have been otherwise at a time when the argument in favor of the Copernican theory was still inconclusive, and when science could offer no more than a few bits and pieces of intellectually interesting astronomical and mechanical theories and the unbounded confidence of Galileo's prophetic genius. In contrast to natural science was the vast body of learning, wisdom, and beauty represented by contemporary humanism and theology. As long as those who had to be convinced by the proponents of science were men who believed in the main depositories of traditional learning, the scientific movement was doomed to failure. For those in the upper class circles who mattered, and probably for the majority of those who did not, science was an intellectually and aesthetically second rate activity as well as a morally and religiously potentially dangerous one. If taken up by an extraordinary genius like Galileo, who could write about science in accomplished literary form, it was given all the attention of a great piece of literature. And if in addition the scientist was a man who could be consulted on great engineering and architectural projects and could show his brilliance in other serious and playful ways, he was honored as a man of outstanding imaginative talents. The term "virtuoso" truly reflects these attitudes and shows the limit of the appreciation of science in Italian society in the seventeenth century.[34]

This attitude to science was not unique to Italy and, had the fate of science depended everywhere in Europe on the same educated and "responsible" ruling class as it did in Italy, the emergence of a proud and self-confident body of scientists might have been postponed for a very long period of time, perhaps indefinitely. But fortunately for the development of science, the social structure in Northern Europe was different. As has been said, there existed in Northern Europe a mobile class whose aspirations, beliefs, and interests—intellectually as well as economically and socially—were well served by their affirmation of the

[34] Ludovico Geymonat, *Galileo Galilei* (New York: McGraw-Hill, 1965), pp. 136–155; Olschki, *op. cit.*, Vol. III, p. 118; Giorgio de Santillana, *The Crime of Galileo* (Chicago: University of Chicago Press, 1955), pp. 104–106.

utopian claims made on behalf of science. Furthermore, part of this class found science a religiously more acceptable intellectual pursuit than traditional philosophy. Thus when the ebbing tide of science, which was receding from the scientific circles and academies of Italy, finally touched France and England, its direction was reversed. The changes that took place at that time set in motion a flood which has still to cease.

The Higher Evaluation of Science in Northern Europe

The most obvious aspect of the transformation that occurred in the scientific movement in Northern Europe was that there science eventually became a central element in an emerging conception of progress. This evaluation was not at all clear from the beginning, however, and many aspects of the movement there appeared to be no more than a reproduction of Italian patterns. The rapprochement between artists and practical men on the one hand and scholars of scientific bent on the other, such as existed in Italy from the fifteenth century on, was copied in other parts of Europe in the sixteenth century. The best-known names are Vesalius, Dürer, and Christopher Wren. The last, one of the greatest architects of the seventeenth century, can be seen as a later and more advanced version than his fifteenth-century Italian forerunners, Alberti and Brunelleschi. Similarly, the northern scientific academies owed their inspiration to Italy. Peiresc, the originator of the informal circles from which the Académie des Sciences eventually arose, was a student at Padua, a correspondent of Galileo, and a disciple of della Porta (who had founded one of the early Italian scientific academies in Naples). He became the center of a continent-wide circle of scientific and scholarly correspondents and visitors; there is a direct link between this circle and those which advocated the establishment of the Royal Society and the Académie des Sciences. But Peiresc only continued what had begun with Galileo, who himself had been the center of a network of correspondents and visitors.[35]

Nonetheless, as early as the sixteenth century the differences between the Northern European and the Italian patterns became evident in a variety of forms. The most important network of scientists and practical men was that concerned with navigation in England and Holland. In England this group included the mathematicians Robert Recorde (1510–1558) and John Dee (1527–1606), both of whom served as consultants to large trading companies. Dee was also adviser to such famous seafarers as Martin Frobisher, Sir Humphrey Gilbert, John Davis, and Sir Walter Raleigh. Thomas Digges, the astronomer who advanced Copernicus' ideas an important step further, also spent some time at sea and interested himself in navigation. Henry Biggs (1561–1630), the first professor of mathematics at Gresham College in London, was a member of the

[35] The central place of Galileo and of Italy is shown by the fact that Peiresc and other Western scholars wrote to him in Italian. See Olschki, *op. cit.*, Vol. III, pp. 440–445.

the emergence of the scientific role

London (later the Virginia) Company, the pioneer group that sailed to the New World. Gilbert's famous treatise on magnetism used the observations of the seafarers Robert Norman and William Borough. The seventeenth-century Cambridge anti-Aristotelian author William Watts used the observations of another seafarer, Thomas James. Richard Norwood, the London mathematician, surveyed the Bermudas for the Bermuda Company.[36]

Associations between scientists and practical men were not confined to matters connected with navigation. Apart from the already mentioned relationship with artists and engineers, there was increasing interest in machines, mining, lens grinding, and the making of watches and other instruments. In contrast to Italy, the germane subjects between scholars and practical people shifted from art and civil and military engineering, which were primarily the concerns of the ruling and aristocratic classes, to navigation and instrument making. These latter fields were closely tied to the concerns and the fortunes of a new, increasingly numerous and self-esteeming class of sea traders, merchants, and artisans. Some of these artisans were also dependent primarily on sea trade.[37] Compared with the social contacts of science in sixteenth-century Italy, these were relatively humble connections. The merchants and artisans were ascending in status and influence, but they still had a very long way to rise.[38] Their status was comparable to the status of the fifteenth-century Italian artist-engineers who at that time had cultivated the company of the scientists.

Potentially, however, this was a much more promising social base for sciences than had ever existed in Italy. The Italian artist-engineers had been dependent for their income on the ruling families who were the exclusive customers for the goods and services produced. These ruling families constituted a small and closed group of people. The change of its composition, due to the absorption of large merchant families into this class, was not enough to change the aristocratic character of this group and the hierarchic character of society as a whole. The cities had continued to be small and closed political units composed of guilds carefully isolated from each other and graded by legal privilege and traditional values. The political units were topped by a ruling class with privileges and power transcending those of the particular guilds. The relationship of the city to its near and far environs had not changed either. It constituted an isle of particular privileges and traditions competing with similar units of privilege and tradition for the rule over the nearby exploitable agricultural

[36] Richard Foster Jones, *Ancients and Moderns* (Berkeley and Los Angeles: University of California Press, 1965), pp. 75–77; Christopher Hill, *Intellectual Origins of the English Revolution* (Oxford: Clarendon Press, 1965), pp. 14–130; Rossi, *op. cit.*, pp. 13–14, 18–19.

[37] There is no evidence about close contact between scientists and this new class of people in France, but the class itself existed there, too, and was gaining in strength and wealth. Italy was at that time rapidly losing its position as a seatrading nation. See F. L. Carsten, "The Age of Louis XIV," in *New Cambridge Modern History*, Vol. V (Cambridge: Cambridge University Press, 1958), pp. 27–30; C. M. Cipolla, *loc. cit.*; F. C. Spooner, "The Reformation in Difficulties: France, 1519–1559," in *New Cambridge Modern History*, Vol. II, pp. 210–226.

[38] Lawrence Stone, *The Crisis of the English Aristocracy, 1558–1641* (Oxford: Clarendon Press, 1965), pp. 21–53.

the emergence of the scientific role

population and over the ancient routes of maritime trades in the Mediterranean.[39]

The scientists, like the artist-engineers, had to find their place within this hierarchy. There was nowhere to go without it. The scientists' only chance of exerting influence and attaining high prestige was to move up into the aristocratic classes.

There was a different world in Western Europe. Trade was expanding beyond boundaries that anyone had ever imagined. As a result, those in the cities and the trading and artisan classes grew beyond the limits of the guild.[40] However, this growth had not changed either the structure or the conception of the class system until the middle of the seventeenth century. The aristocracy remained the only class with universal influence and prestige.[41] But important sections of this class participated in the trading activity. They could not fail to see the new perspective of an expanding economy and a much more open class system with variable and changing ways of life. Or, where the aristocracy had been cut off from the new developments—as it had been in France—there was a king with his advisers to take cognizance of it.

Although this was the general situation, it does not follow that the ruling classes, or people in general, had been more progessive in the West than in Italy. Science—potentially subversive to religious authority and technologically of limited importance—merited low priority everywhere, and those responsible for law and order could give it only limited and qualified recognition. Therefore, there was no question anywhere in the sixteenth or seventeenth century of the official and general acceptance of a science-centered philosophy. To the extent that such ideas occurred at all—though unsuccessfully—they probably happened most often in Italy because it had a broader stratum of highly educated people than any other country.

The development of science depended on the determination of the minority who believed in science to fight for its general recognition openly and to express and develop its interest in science in public discussion and purposeful association. The growth of science in the sixteenth and most of the seventeenth century thus depended on the existence of some plurality of cultural interests and of scales of social evaluations. The extent to which these varying interests and evaluations were allowed was a function of the openness—or, by the standards of that time, imperfections—of the class system. Where there were individuals and groups whose rapidly rising fortunes came from the discovery of new places, new routes, new markets and new kinds of goods, there was greater readiness to consider the claims of science as a more valid way to truth than traditional philosophy. The recognition came about partly, perhaps, because these claims

[39] Cipolla, op. cit., pp. 397–430.
[40] Pieter Geyl, Revolt of the Netherlands, 1555–1609 (London: Williams and Norgate Ltd., 1945), pp. 38–44; H. Koenigsberger, "The Empire of Charles V in Europe," New Cambridge Modern History, Vol. II, pp. 301–334; G. Spini, "Italy After the 30-years War," New Cambridge Modern History, Vol. V, pp. 458–473.
[41] Stone, op. cit., pp. 39–44.

the emergence of the scientific role

fitted the new perspective of a socially and materially changing world well; but it also came about more decisively because the interests of these groups were opposed to oppressive assertions of traditional privilege in general.

The Religious Factor and the Rise
of a Scientific Utopia

The other important condition that increased the likelihood of recognition of an autonomous scientific outlook in the West rather than in Italy was the difference in the religious situation. Man lives not by bread only but also by word of God, and this was particularly true in the seventeenth century. Nearly everyone in Europe was religious, either Christian or Jewish. The church had managed to come to terms with philosophies that opposed its doctrine more directly than science. But it was easier to allow free speculation about abstract things such as the immortality of the soul than to submit specific questions like the nature of the moon to the test of the telescope. The speculations of the human mind about religious questions could never be conclusive. In questions in which speculation was regarded as the right method, it was a matter of God's power versus man's mind. God's ultimate power was beyond the power of man's mind, and where there was a contradiction between the divine mind and human minds it was not difficult to "see" where the ultimate truth lay. But empirical science—once it touched on questions of basic theological importance—did not allow such evasion; it confronted empirically observed nature as *actually* created by God (in the view of practically everyone at that time) to written records that were authoritatively accepted as His own words or directly inspired by Him. The discrepancies between the empirical observations and the authoritative records became more and more evident. As a result, Catholic, Protestant, or Jewish religious authorities tended to take an attitude ranging from hostility to extreme caution with respect to empirical science.

Among the major European religions, however, there was one important difference: Protestantism did not possess a universally constituted religious authority, and its doctrines left the interpretation of the Bible to the individual believer and permitted him to seek his own religious enlightenment. A Catholic or a Jew had to suppress what might be his conviction that science would ultimately prove to be a new way to God because of his religion's fixity of biblical interpretation. However, a Protestant who felt that God's will and the discoveries of science were in harmony could speak in good conscience, provided that he lived in an environment where church authority was unstable or weak. (Where the authority of the clergymen was well established, it was their usually antiscientific interpretations that prevailed.)

Thus the idea that science and technology (the "practical arts") could become a better mode of education and an improved intellectual and moral culture was consistent with the interests and the outlook of the mobile middle classes in general. However, only certain branches of Protestants could make

69

scientific knowledge (or a philosophy granting such knowledge complete autonomy), an integral part of their religious beliefs. Only they could thereby overcome the resistance that religious belief might interpose. Protestantism thus provided the legitimation for a new utopian world view where science, experiment, and experience were to form the core of a new culture even though the logical relationship among the three was perhaps mistakenly construed.

The beginnings of the ideas that link science, the practical arts, and the continuous improvement of man's condition can be traced back to the 1530s. Ludovico Vives, a Spanish Protestant scholar who was tutor at the English court, was among the first to extol the educational and intellectual virtue of the experience of the artisans.[42] Starting from the middle of the sixteenth century, however, these Renaissance beginnings were taken up by Protestant philosophers and educationists and transformed into what Karl Mannheim called a "utopia." The originators of this trend were Peter Ramus and Bernard Palissy, and they were followed by Francis Bacon, Comenius, Samuel Hartlib, and several others. They were interested in universal education and far-reaching projects of scientific and technological cooperation, which they hoped would lead to the conquest of nature and the emergence of a new civilization. They believed in a worldly redemption to which science and technology and their effective support and organization would lead.[43]

None of these people was a significant scientist, nor—with the ambiguous exception of Bacon—even an important philosopher. They were publicists who were interested in practical results; they expressed programmatically the intellectual outlook of the circles of scientists and others who cooperated in the solution of practical problems. In Italy this cooperation had never been turned into an outlook with primarily practical aims of social reform. The only attempt with implications for wider issues was that of Galileo, which ended in failure. But even his aim was the conversion of the church to the right cosmological beliefs and the modernization of the intellectual life of Italy; it was not social improvement. The fact that science was turned into such a broad practical outlook in Northern Europe reflected the beginnings of an open system of classes; that this outlook was taken up by intellectuals and developed into an ideology, potentially threatening to traditional authority was possible only as a result of the doctrinal fluidity of Protestantism.

---

[42] Other pioneers of the idea, like Rabelais, Montaigne, Erasmus, were Catholics. Rabelais might have had Italian models before his eyes. The education of Gargantua, which reflects his ideas on the subject, is a university education combined with the typical training of an Italian artist of the fifteenth century. Alberti had had this type of education and the training provided later in the Accademia del Disegno—one of whose teachers, Ostillio Ricci, was Galileo's private tutor in Florence—(founded 1563) was of this Gargantuan kind, if not of similar dimensions. About the spread of these ideas all over Europe during the sixteenth century and the educational experiments inspired by it see Rossi, *op. cit.*, pp. 15–16. The classic treatment of the whole web of relationship between Protestantism and science is R. R. Merton, "Science, Technology, and Society in Seventeenth Century England," *Osiris* (1938), IV: 360–632.

[43] Jones, *op. cit.*, pp. 62–180.

70

Protestant Policy on Science

It was not that all varieties of Protestantism adopted this new view of science or that they did so regardless of their location. In small, self-contained Protestant communities, such as those in Geneva and in Scotland, in most places in Germany and, later in the seventeenth century in Holland, science fared worse than in the great Catholic centers of Italy, France, and Central Europe. These Protestant communities were small and tightly knit; because they were relatively undifferentiated, they had no appreciable class of intellectuals except their clergymen.[44] Like the similarly organized Jewish communities, they would not tolerate anything approaching heresy. The Catholic Church, on the other hand, with its tradition of learning and its own large class of differentiated intellectuals in the teaching orders usually had more sympathy for specialized, nonreligious intellectual interests.

In most places, however, Protestants were unable to form a closed religious community. On the one hand, they were in contention with the Catholics; on the other hand, the various Protestant sects fought among themselves. In those situations, no effective religious authority existed to enforce conformity in doctrine and practice. The governments of the territories where these conditions obtained were much freer than any others to adopt a sympathetic attitude to science and the scientistic utopia. Those who believed in the utopia were free to propagate their views, and the official authorities could adopt a pragmatic attitude toward the matter. As a result official Protestant authorities on several occasions adopted policies of supporting science, and eventually in Commonwealth England they came very near to the acceptance of the scientistic utopia as a basis for their official educational policy.

The first notable opportunity for the emergence of a distinctly positive Protestant policy on science was provided by the persecution of Galileo. Any oppressive act by the Catholic Church, which was their greatest opponent in religious competition, was immediately used as propaganda. The case of Galileo was a conspicuous one. Immediately after his condemnation, a group of Protestant scholars in Paris, Strasbourg, Heidelberg, and Tübingen decided to translate Galileo's work into Latin. In this endeavor they received general support from several Protestant communities otherwise not notable for their tolerance of Copernican ideas. Copies of the original work were obtained through doctrinally rigid Geneva; one member of the group was from Tübingen University, where only some time before, Kepler had been prevented from earning a theological degree because of his Copernican views.

The Dutch Government also turned Galileo's disfavor to Protestant advantage by inviting him, through Grotius, to advise it on the measurement of longitudes. Even though Galileo's advice was not followed, official honors were bestowed on him by the Dutch Government, and the communications continued until they were interrupted by the Curia, which—probably not incorrectly

---

[44] A. de Condolle, *Histoire des Sciences et des Savants*, 2nd ed. (Geneva: H. Georg, 1885), pp. 335–336.

—perceived that they were being exploited for purposes of Protestant propaganda.[45]

Protestant scholars with scientific interests saw Galileo's persecution as an opportunity to connect the furtherance of the Protestant cause with achievement of official support for science. Their concerted action was perhaps the earliest manifestation of an active scientific lobby in Protestant Europe. At least some of the intellectuals involved were acting on behalf of the promotion of science and not just for a general, religious-educational cause.

It is difficult to conclude just how long the Galileo case was used to link science with Protestantism. In any case, it was not a major factor in the establishment of science. In England science became involved in Protestant politics in a new and more significant way. Both before and under the Commonwealth, it had become increasingly difficult to maintain public consensus on anything of potential religious importance because of the numerous theological dissensions with political implications. One of the often-mentioned features of the prehistory of the Royal Society is that the participants at the informal meetings of the circle from which the society emerged agreed not to discuss matters of religion and politics but to restrict themselves dispassionately to the neutral field of science.[46] For apparently similar reasons, the Baconian philosophy and the support of science became part of the official policy of the Commonwealth. One of the educational-scientific publicists of Commonwealth England was John Durie, who had spent much of his time in Northern Europe working on the unification of all the evangelical churches. Both Hartlib, who supported Durie, and Haak, another member of the early group of scientists and politicians who promoted science, were probably similarly motivated by their personal experiences with religious conflicts in Europe. They were influential in the formation of educational policies for the Commonwealth, and their ideas became official policy. The sudden rise of the popularity of the Baconian view in the late 1640s, and the appointment of Wilkins, Wallis, Petty, and Goddard to university chairs testified to the success of the new ideas.[47] In addition to its congeniality to the class interests of artisans, merchants, and other mobile people who formed the backbone of the regime Baconian science was something on which the more enlightened elements among the Puritans could agree. Scientific activity was welcomed by those who were interested in a more secular education and who shared the distaste of anything that reminded them of the old regime. Yet it was also acceptable to the more fanatical Puritans who thought that the study of the Bible should be sufficient education for all and even wanted to abolish the universities altogether, since science appeared to them as a lesser evil than pagan humanistic learning. Science thereby found hospitality in its own right. It did so not because it was positively supported by any particular doctrine of

[45] Olschki, op. cit., Vol. III, pp. 401–403, 440–443.
[46] T. Sprat, History of the Royal Society of London (London: J. R. for J. Martyn, 1667).
[47] Jones, op. cit., pp. 109, 117.

72

Protestant theology, but because it was relatively free from involvement in the theological and philosophical disputes that had ravaged the Continent and which were disrupting English society as well.[48]

The implications of these circumstances are crucial for an understanding of the rise of modern science. The scientistic world view—as distinct from actual science—had not been adopted because it offered a better philosophy or a better explanation of important natural phenomena than the previous philosophies and religious doctrines. Those who were satisfied with the world as it was did not change their scale of intellectual values as a result of the better solutions to a few riddles of nature. But for those interested in changing the world, empirical science was the true prophecy. It produced innovations that contained their own incontrovertible proofs and made all philosophical controversy unnecessary. And not only was it a way to innovation, but also to social peace, as it made possible agreement concerning research procedures to specific problems without requiring agreement on anything else.

This acceptance also explains the otherwise somewhat baffling attitude of reverence toward Baconian experimentalism in seventeenth century English science. Bacon was a bad scientist, and in many details he was not a very good philosopher either. There was little connection between the rise of new astronomy and mathematical physics and Baconian principles; experimentation without theory and collection of empirical knowledge had produced few scientific results.

Without an agreement on the experimental method, however, an autonomous scientific community could never have arisen. Had science been presented as a superior but logically closed and coherent philosophy, it would have become one of the warring philosophies rather than a neutral meeting ground. Even if by some unlikely chance it had been made into official philosophy by some ruler, it would have soon been turned into a comprehensive and diffuse philosophy. This actually happened to Cartesianism. And in the eighteenth century there arose a strong tendency to apotheosize even Newton and turn him into the pivot of a new comprehensive and essentially static philosophy. There would have been little resistance to this tendency by the scientifically erudite. As a matter of fact they would have participated in the effort, as Leibniz or Christian von Wolfe did.[49]

Baconism, however, opposed such a closure of the scientific outlook by creating the blueprint of an ever-expanding and changing, yet regularly functioning scientific community. The experimental doctrine was not a theory, but it was a valid strategy of conduct for scientists. For those who adopted it, it

[48] Although I know of no evidence for it, it is not impossible that the feeling that science was above and beyond the theological and philosophical disputes so discredited by religious strife had something to do with Boyle's and Newton's belief that it might constitute a new way to God.

[49] For the interpretation of Bacon as a strategist of empirical science, see Margery Purver, *The Royal Society: Concept and Creation* (London: Routledge and Kegan Paul, 1967), pp. 20–100. But she seems to imply that this interpretation is inconsistent with an emphasis on the relations of the scientists with the scientistic movement.

became a medium of unequivocal communication, a way of reasoning and of refutation in limited fields of common interest. By sticking to empirically verified facts (preferably by controlled experiment), the method enabled its practitioners to feel like members of the same "community," even in the absence of a commonly accepted theory. It was possible for scientists to go ahead with several competing views of the common subject matter and have the feeling of shared progress and eventual consensus. They no longer had to split into factions opposing each other on an increasingly wide and diffuse front as the case had been before in philosophical conflicts.

Under the conditions of ideological impasse that were reached in England by 1640, scientists found themselves in a situation in which it was increasingly useful to adopt Baconianism as a strategy of survival within and outside science. Thus natural science became the paradigm for the philosophy of an open and plural society. During the crucial period from the Puritan Revolution to the Glorious Revolution, natural science served as the symbol of a neutral meeting ground for the useful pursuit of common intellectual goals. Its method of trial and error (or hypotheses and its refutation) created a time perspective that made possible acceptance of situations of imperfect knowledge and the absence of a consensus by requiring a consensus on procedures only. Science that was considered Baconian, was taken as proof that consensus on procedures would eventually produce valid results.[50]

This acceptance of Baconianism explains why there was more sustained social support for the scientistic movement in the West than in Italy, in spite of the cultural superiority of the latter country during the sixteenth and most of the seventeenth century. It also explains why revolutionary England, of all the countries of the West, became the center of the movement during the middle of the seventeenth century and how it happened that in England the natural scientists, who had been a rising faction in philosophy since the fifteenth century, became an autonomous, distinct, and respected intellectual community.

[50] That, indeed, the scientific method was considered as a paradigm of arriving at a consensus in an objective and impersonal manner can be seen from its use in economic and political theory. See William Letwin, *The Origins of Scientific Economics* (Garden City: Anchor Books, Doubleday, 1965), pp. 131–138.

the emergence of the scientific role

the institutionalization of science in seventeenth century england

five

The crucial difference in the place of science in England as compared with other countries about 1700 was that in England science was institutionalized. As the words "institution" and "institutionalization" are used in a variety of ways, the terms need definition. Here institutionalization will mean (1) the acceptance in a society of a certain activity as an important social function valued for its own sake; (2) the existence of norms that regulate conduct in the given field of activity in a manner consistent with the realization of its aims and autonomy from other activities; and finally, (3) some adaptation of social norms in other fields of activity to the norms of the given activity. A social institution is an activity that has been so institutionalized.[1]

In the case of science, institutionalization implies the recognition of exact and empirical research as a method of inquiry that leads to the discovery of important new knowledge. Such knowledge is distinct and independent of other ways of acquiring knowledge such as tradition, speculation, or revelation. It imposes certain moral obligations on its practitioners: completely universalistic

[1] This definition of the word "institution" is closely related to that of S. N. Eisenstadt, "Social Institutions," *International Encyclopedia of Social Science*, Vol. 14, pp. 409–410. It is to be distinguished from that usage, which also includes in "institution" the actual organization of social activity in a given field. The autonomy of values and the norms of each social institution are emphasized by Norman W. Storer, in *The Social System of Science* (New York: Holt, Rinehart & Winston, Inc., 1966), pp. 39, 55–56 (Storer speaks about a "social system" to describe what I call "institution").

evaluation of contributions; the obligation to communicate one's discoveries to the public for use and criticism; the proper acknowledgment of the contributions of others; and, finally a variety of conditions in other institutionalized fields, such as freedom of speech and publication, a measure of religious and political tolerance (otherwise it is difficult to maintain universalism), and a certain flexibility to make society and culture adaptable to constant change that results from the freedom of inquiry.[2]

The independence of science from other fields of inquiry and the recognition of the norms of science as independent from other norms were part of the official program of the Royal Society. That independence was also manifested in the work style of the members of the Royal Society. Unlike their great continental counterparts, such as Descartes, Gassendi, and Leibniz, the members of the Royal Society did not consider their scientific work as part of a broader speculative philosophy. Usually they did not engage in such activity at all, as they considered empirical science an occupation of sufficient or even superior dignity in its own right.[3]

Another manifestation of the greater autonomy of science from traditional philosophy in England was the battle between the "ancients" and the "moderns" in the different countries. On the Continent the "moderns" still had to fight for equality against official theological and traditional philosophical authorities, while in England by the end of the seventeenth century the situation was already ripe for an intellectual counterattack against the naïve and excessive claims of the scientists and their followers.[4]

Finally, only in England was there a significant adaptation of institutional norms in general to the requirements of autonomous science. As shown in the previous chapter, the rise of the scientific movement in England was tied from the very beginning to the rise of religious pluralism and social change. The importance of science in the educational philosophies and policies of the Cromwell era has also been pointed out. These ideas survived and had an influence on the academies of the dissenters.[5] Finally, starting from the 1660s, there emerged a series of attempts to shape political and economic philosophy and practice according to the model of self-regulating mechanical systems, rather than as an

---

[2] About these characteristics of science see Robert K. Merton, *Social Theory and Social Structure*, rev. ed. (New York: The Free Press, 1957), pp. 550–561; Bernard Barber, *Science and the Social Order* (New York: The Free Press, 1952), pp. 122–142; and Storer, *op. cit.*, pp. 75–90.

[3] Dorothy Stimson, *Scientists and Amateurs* (London: Schuman, 1948), pp. 73–76; M. Purver, *The Royal Society: Concept and Creation* (London: Routledge & Kegan Paul, 1967), pp. 34–64, 111; Alexandre Koyré, *From the Closed World to Infinite Universe* (New York, Harper Torch Books, 1958), pp. 159–160.

[4] Stimson, *op. cit.*, pp. 70 ff.; and Richard F. Jones, *op. cit.*, pp. 237–272.

[5] Irene Parker, *Dissenting Academies in England: Their Rise and Progress and Their Place Among the Educational Systems of the Country* (Cambridge: Cambridge University Press, 1914), pp. 41–49.

the institutionalization of science in seventeenth century england

order imposed by supreme authority.[6] Thus political society was conceived as composed of independent individuals, as matter was composed of atoms, and as held in balance by the conflicting forces of the executive, the legislature, and the vested interests of the crown, the aristocracy, and the corporations. Economic theory dealt with quantities, such as supply and demand, and the amount of money and their equilibrium reflected in price. On the Continent, on the other hand, science was still considered a potentially dangerous and subversive philosophy whose influence on political, economic, and religious-moral behavior had to be closely checked and curtailed.

These are of course only very gross outlines of the situation. Active interest in science was limited to a few individuals, and passive interest in science was also limited to small and largely upper class groups everywhere. But in England these groups became irreversibly part of official society by the end of the seventeenth century. It was possible and not very dangerous to try to apply the new experimental approach to practically any part of private or public life. Elsewhere, supporters of science and scientistic philosophy had to confine their support to pure science and technology. Any extension of the scientific approach to public matters was wrought with danger of prosecution by church and government.

The Shift of Interest from Science to Philosophy and Technology

Paradoxically, the institutionalization of science did not have the effect of maintaining the scientific leadership of England. During the eighteenth century the Royal Society became a club of amateur philosophers and naturalists, and eventually it was eclipsed by the French *Académie des sciences* as the world's leading scientific society.[7]

This is not to say that there had been an actual decline of science in Britain. Rather, scientific activity had become dispersed and had lost its center somewhat. But even this observation is only partially true, since there arose a new center of science in the Scottish universities during the second half of the eighteenth century. There is also some evidence about the existence of great

---

[6] About the influence of the scientific model on Locke's economic theories see William Letwin, *The Origins of Scientific Economics* (Garden City: Doubleday Anchor Books, 1965), pp. 176–178; for a more general discussion of this aspect of Locke's philosophy see Charles C. Gillispie, *The Edge of Objectivity: An Essay on the History of Scientific Ideas* (Princeton: Princeton University Press, 1960), pp. 159–164.

[7] About the decline of the Royal Society in the course of the eighteenth century see Stimson, *op. cit.*, p. 140. According to the author there was not a single scientist among the members of the Royal Society in ten different years during the eighteenth century. The fellows were mainly antiquarians, historians, and librarians. About the central importance of France and the *Académie des Sciences* see John Theodore Merz, A *History of European Thought in the 19th Century* (New York: Dover Publications, Inc., 1965), Vol. I, pp. 41, 89–109.

77

popular interest in science in England.[8] And the influence of scientific thinking, or at least of natural science as a model for valid thinking about political, economic, and technological matters and practice was more widespread in Britain than anywhere else.

Nevertheless, in the course of the eighteenth century, France became the center of world science. By the last decades of the century, the quality of French science surpassed that of the British in every field. The *Académie des sciences* became the world's most prestigious scientific organization. Advanced students from all the countries of Europe went to Paris to acquaint themselves with the most recent developments of science and French was used as the *lingua franca* of scientists and scientific bodies throughout Europe.[9]

It seems a paradox that just when science became institutionalized in England that country lost its leadership in science to France, which was a much more traditional society. In order to explain this development, the relationship between the scientistic movements and institutionalized science has to be clarified. Neither scientistic movement nor institutionalization of science refers to expert scientists or expert scientific activity, but to the behavior of people in general in relation to science. The scientistic movement consists of a group of people who believe in science (even though they may not understand it) as a valid way to truth and to effective mastery over nature as well as to the solution of the problems of the individual and his society. Empirical and mathematical science, in this view, is a model for the solution of problems in general and a symbol of the infinite perfectability of the world.[10] The word "movement" implies that the group strives to spread its views and to make them acceptable to society as a whole. Institutionalization follows when the movement achieves its aim and has its values actually adopted by society.

The explanation of the relative decline of science in eighteenth century England, and its relatively rapid rise in France seems to be in the difference between the movement and institutional stages of science. The hypothesis I suggest is that at the movement stage an additional social motivation is generated which becomes diffused and hence loses intensity during the institutional stage. Thus toward the middle of the seventeenth century, science became an important and perhaps central symbol of an open and advancing society which was the ideal of powerful social groups in England. These groups, however, were still a minority in conflict with a largely traditional "official" society. Neither their beliefs nor their purposes were shared by the majority, or at least by the

---

[8] Nicholas Hans, *New Trends of Education in the 18th Century* (London: Routledge & Kegan Paul, 1951), pp. 155–158.

[9] Merz, *op. cit., loc. cit.*

[10] The term "scientistic movement" is preferred to that of "scientific movement" used by Jones, *op. cit.* in his description of the rise of Baconianism because of the importance of distinguishing between expert scientists and the closely related movement for which Baconian philosophy and experimental science were general principles to be applied to all human and social problems. About the development of this movement after the age of Newton and Locke and its links with French Enlightenment see Gillispie, *op. cit.*, pp. 151–178.

the institutionalization of science in seventeenth century england

majority of those who mattered. They therefore had few opportunities to try to put their ideas to the test of reality as far as social or political reforms were concerned.

As long as this situation lasted, there was a steeply rising interest in science. For a gifted few, science served as a haven of freedom of thought, speech, and creation in a society where freedom was either suppressed or made meaningless by the absence of consensus on basic religious and political premises. For a much broader group (such as the intellectuals of the scientistic movement who adopted the philosophy of Bacon and those of all classes who followed them), empirical science symbolized a goal that was still to be attained: the creation of a new social order where things could be changed and improved by rational and objective procedures and without constant violent conflict.

Following the Glorious Revolution, the situation changed. The intelligent discourse about the pressing matters of moral, political, and economic philosophy could be resumed and had to be resumed. Since the utopia of an open and pluralistic society was partially attained (at least to the extent that there were no important groups left who felt themselves excluded and badly frustrated), it was time to try and realize the Baconian promises of "advancement."

It was inevitable, therefore, that people turned to social philosophy and technology. They did so with natural science and experimental procedure serving as their guides. But natural science could not provide more than very general guidelines toward the creation of a new social philosophy and not much more toward the solution of technological problems. Therefore philosophers, economists, technologists, and physicians had to pursue their respective interests in largely empirical ways. The few attempts to do it more systematically and theoretically, such as the attempt to create a medical physics (iatrophysics) had failed.[11] This failure explains the shifting of interest to these practical fields and the apparent loss of interest in science.

It has to be emphasized, however, that this loss of interest was more apparent than real. The disillusioning sterility of theoretical science in eighteenth century England did not diminish the belief in the experimental method as the principal means in man's effort to understand and master his physical and social environment. This belief served as the basis of most of these philosophical and technological enterprises. Only the mood of unrealistic expectations from science had passed. Gifted and creative individuals now had a much greater range of opportunities to exercise their creativity than before. It was possible to discuss matters of politics, economy, and philosophy without fear of violent conflict, and there was a fair scope for actually influencing policy. The scientistic promise of technological advance was also put to the test of reality. Experimentation with steam engines, textile machinery, and other technological schemes began and was spurred on by enterprise that flourished under policies that were con-

11 Richard Harrison Shryock, *The Development of Modern Medicine: An Interpretation of the Social and Scientific Factors Involved* (London: Victor Gollancz, Ltd., 1948), pp. 20–40.

the institutionalization of science in seventeenth century england

stantly scrutinized by economists and changed at times as a result of their recommendations. This experimentation was part of the institutionalization of science: a process of trial and error to find the limits of the applicability of scientific principles and to change social institutions according to those principles.

But all these changes implied that (*a*) socially generated motivation was less likely to channel creative talent into natural science than before; and (*b*) advances in pure science lost their rhetorical importance in the public debate over political reform and economic advancement.[12] The basic reforms were made, and the problem was an empirical and not a rhetorical one. Scientific achievements, therefore, had as much value as they were worth to the few practitioners of science who used them and the amateurs who enjoyed them. At the movement stage they also had the added values of having been the only field of free intellectual creativity and of having constituted the strongest argument for liberalism in a basic ideological debate. An abstract way of summarizing would be that at the movement stage a great deal of intellectual stimulus and social interest in all kinds of other activities (religion, economics, politics, etc.) was channeled into science. On the other hand, at the stage of institutionalization, much intellectual stimulation generated by science came to be dispersed over a wide range of activities to which science was applied. As a result, there might have been a relative decline of interest in science as narrowly defined, although the decline was probably more than counterbalanced by the application of a scientific approach to all kinds of other activities.

Scientism and Science in Seventeenth Century France

This explanation also fits the French case. The establishment of the *Académie des sciences* in 1666 in Paris, though prompted by the British example, was no imitation of it. The informal circles and academies in France, which had preceded the *Académie's* establishment and had conducted the propaganda for it, were similar to those in England. As a matter of fact, these informal groups had been in close contact with the English groups as well as with the Italian ones.[13] In the 1630s Marin Mersenne was the central figure for correspondence encompassing all the known scientists of Europe, and he assembled in conferences at his home the leading French scientists of the time: Descartes, Desargues, Gassendi, the brothers Pascal and Roberval. Both Mersenne and Theophraste Renaudot were in touch with Haak and Hartlib in England and were in favor of their and Comenius' ideas on educational reform.

---

[12] This loss of added symbolic value was also manifested in the abatement of the attempt to extend the authority of science over literature and humanistic education and is described, but differently interpreted by Jones, *op. cit.*, pp. 271–272.

[13] About the various groups that preceded the establishment of the *Académie des Sciences* see Harcourt Brown, *Scientific Organizations in Seventeenth Century France (1620–1680)* (Baltimore: The Williams & Wilkins Company, 1934), pp. 2–7, 18–27, 32, 62–66, 75–76, 117–127, 195–199.

the institutionalization of science in seventeenth century england

Renaudot was of Protestant origin—he had been converted to Catholicism after the fall of La Rochelle in 1629—and Mersenne, though in holy orders, was suspected of reformist tendencies. However, there were no Protestants among the scientists of Mersenne's circle.[14]

Of the other politicians of science active in scientific salons including Montmor, Auzout, Hedelin, Thevenot, Petit, and Sorbière, only Sorbière was a Protestant by origin. But they were all greatly influenced by the English example and by Baconian philosophy. The first editor of the *Journal des savants*, Denis de Sallo, was dismissed under clerical pressure because of his pronounced Gallican and Jansenist leanings (both of the latter were doctrinally critical, antiecclesiastical positions). The most important supporters of these groups in the royal court were Richelieu and, later, Colbert.[15]

Thus, among the groups favoring a policy of active support of science and the adoption of the new scientific philosophy in France, Protestants, Gallicans, and Jansenists had played an important role. Like their opposite numbers in Protestant London, they too lived in a situation in which there was no single, well-established church authority. They had an important interest in the recognition of science as an autonomous and theologically neutral field of intellectual activity, since such recognition was to strengthen the cause of religious pluralism, on which the survival of these groups depended.

But the structure of society in France differed from that in England. There was, it is true, a middle class in France which had somewhat similar sources of income as that in England. It, too, was growing and expansive in outlook. But class differences in France were much more rigid, and the power of the king was much more pervasive than in England. Therefore such an attempt at educational and social reform as that advocated by Renaudot did not become the concern of any of the more important Paris circles of scientists and wealthy

[14] About the religious affiliations and attitudes of some of those involved in the French movement, see *ibid.*, p. 36, which quotes a letter about the death of Mersenne where he is referred to as a *Moine Huguenot*. The writer goes on to say: . . . *il ne croyoit pas toute sa Religion . . . et il n'oisoit dire souvent son Breviaire, de peur de gater son bon Latin*. (". . . he did not believe all his religion . . . and he did not dare to say his Breviary frequently because he feared to spoil his good Latin.") The writer of the letter, André Pineau, was a Protestant. About the contacts of the French group with Haak in England, see pp. 43–47; for the English contacts and Protestant leadership of the group organized in Caen in 1602, see pp. 216–217; about the Gallican and Jansenist tendencies in the *Journal des savants* under de Sallo, see pp. 193–197. For another important figure in the scientistic movement, consider Henry Justel, who was also a Protestant and became important only in the 1660s (see p. 180). On Renaudot see R. Duplantier, "le vie tourmentée et laborieuse de Theophraste Renaudot," *Bulletin de la Société des Antiquaires de l'Ouest* (Poitiers), XIV, 3rd series, (1947, 3rd and 4th trimesters), pp. 292–331. On Sorbière see *Biographie Universelle, Ancienne et Moderne*, vol. 43 (Paris: Chez L. G. Michaud, 1825), p. 123, and *A Voyage to England Containing Many Things Relating to the State of Learning, Religion and Other Curiosities of the Kingdom*, by Mons. Sorbière; as also *Observations on the Same Voyage*, by Dr. Thomas Sprat, Fellow of the Royal Society, and now Lord Bishop of Rochester, with a Letter of Monsieur Sorbière concerning the war between England and Holland in 1652. To all which is prefixed his Life by M. Graverol, London, T. Woodward, 1709.

[15] About the dismissal of Denis de Sallo, see Harcourt Brown, *op. cit.*, pp. 188–195.

amateurs. (In Loudun, the native city of Renaudot, there was apparently a group interested in science as well as social reform, and it included both Protestants and Catholics).[16] The group immediately preceding the *Académie des sciences*, the so-called Montmor Academy, was much more of an upper class salon than the group from which the Royal Society arose. It was also more difficult in Paris than in London to overcome the resistance of traditional vested interest groups such as the medical faculty, the Sorbonne, the Jesuits, and others.[17]

The result was that those in the "scientistic movement," namely, the group for which science had broad social and technological implications, was not represented in the *Académie des sciences* when it was eventually established in 1666. Whereas the Royal Society was an independent corporation based on mixed membership including amateurs and politicians of science as well as scientists of outstanding accomplishments, the *Académie* was a sort of elevated scientific civil service composed only of a small number of scientists of high reputation.

The intention of this arrangement was to control science and limit its influence to such matters as were judged desirable by French royalty. Royal recognition was given to exact and empirical science and to technology under the condition that the empirical and experimental approach of science would not be diffused to politics and that the norms of universalism of science would not be applied to matters of religion and social estate.[18]

These qualifications were also present in England during the Restoration up until the Glorious Revolution. This situation is evident from the programmatic announcements in Bishop Sprat's *History of the Royal Society*, which emphasizes the value neutrality of science. But the social composition of the society and the activities of some of its members show a close connection with the scientistic movement.[19]

Thus while in England the establishment of the Royal Society was a decisive step toward the institutionalization of science, this was not the case with the French *Académie des sciences*. The former was part of a process that legitimized religious pluralism and social change. Science was recognized as a value in itself, but it was taken for granted that this also implied a measure of recognition of the scientistic movement and a step toward a liberal society. This implication was not officially accepted under the Restoration, but even then it was fairly obvious.

In France, on the other hand, the way the *Académie des sciences* was set up expressed an intention to disrupt the link between expert science and technology and the scientistic movement. It was an attempt to insulate science from the rest of the social institutions. Science was supported under the condition that it would pursue its own aims. In addition, science was to serve the economic

---

16 Duplantier, *op. cit.*

17 Brown, *op. cit.*, pp. 142–148 about the influence of social rigidities and the power of organized vested interests in shaping the development of the scientistic movement in France.

18 *Ibid*, pp. 117–118, 147–148, 160, 200.

19 Purver, *op. cit.*, pp. 34–36, 111.

82

the institutionalization of science in seventeenth century england

and military purposes of the absolutist monarchy, but it was to dissociate itself from broader social movements.

As a result, the scientistic movement continued to persist in France and reached its greatest influence in the Enlightenment movement of the eighteenth century. The dynamics of the situation throughout the eighteenth century were similar to those in England during the Restoration period. The scientistic movement and the groups supporting it were kept apart from political power and frustrated by traditionalist religion. Science was granted qualified recognition by royalty, but it served as a symbol of progress in a much broader sense for a scientistic movement increasingly gaining strength.

Thus science in France continued to retain its symbolical importance for the scientistic movement and its privileged position for intellectually creative people as the only field of free and safe intellectual activity. The *Académie* and its numerous provincial counterparts even possessed the right to publish their proceedings without having to submit them to censorship.[20] The fact that science was supported by an absolutist government did not diminish its value for the scientistic movement. After all this support could be interpreted as another proof of the inevitability of progress.

Thus the explanation given to the rise and relative decline of science in England also explains the shift of the center of worldwide scientific activity to France in the eighteenth century. In England, by that time, science was institutionalized and thus lost the symbolical importance that had been lent to it by the scientistic movement when the latter still strove to realize its aims. In France, however, the full institutionalization of science was prevented. Hence there continued to be an increasingly strong scientistic movement which gave science the added symbolic value that it had lost in England.

A Comparison of the State
of Science in England and France
in the Eighteenth Century

The basic similarity of the elements and the dynamics of the situation in England and France is manifested also in the similarity of the social characteristics and functions of the scientists as compared with other intellectuals in the two countries. Although the *Académie* became a seat of expert scientific investigation, and although there were also a few other government-supported institutions employing scientists in prerevolutionary France (there had been none in Britain), these centers provided scientists with

[20] D. Kronick, *A History of Scientific and Technical Periodicals* (New York: Scarecrow Press, 1962), pp. 132–133. This book deals with the privileges of the *Académie des belles-lettres, sciences, et art* of Rouen which were modeled on those of the *Académie des sciences*, and the *Académies des inscriptions et belles-lettres* in Paris. The privileges were apparently first mentioned in the statutes of the Academy of Caen, *ibid.*, pp. 139–140. In England, where official censorship was not rigorous, scientific societies had only to fear the merciless criticism of newspapers. See *ibid.*, pp. 140–141.

the institutionalization of science in seventeenth century england

positions that were considered as elite, and not as regular, careers. The bulk of the scientists in both countries were amateurs who came from the upper bourgeoisie and the aristocracy who could afford to spend their own time and money on research.[21] The official positions in France were designed to reward a few worthy individuals, and, if they did not come from the wealthy classes, to elevate them into those classes. However, they were not designed as an attempt to transform science into a regular occupation.

The fact that science had the same definition and the same functions in both countries meant that the purpose of empirical natural science and the function of scientists was to make discoveries. This distinguished science from scholarship, which dealt with the restoration, preservation, and transmission of literary tradition. The expected uses of science were also similar in both countries. They were first and foremost technological, which, as it has been pointed out, made science acceptable also to the conservative groups. In addition, there were in both countries groups of intellectuals, so-called philosophers backed by wealthy social classes for whom natural science and mathematics served as the models of correct thinking and as paradigms to deal with social problems. In Britain, these classes participated in the government directly. In France they exerted influence through the intermediacy of some civil servants but were usually frustrated by the government. The informal relationship between expert scientists and philosophers was close in both countries. There was also a distinction in both countries between the expert natural scientists and other philosophers, although this latter distinction was to some extent formalized in France, and, at times, blurred in Britain. Scientists and other philosophers of the two countries were in close intellectual communication with each other, and ideas and discoveries (with the exception of mathematics) traveled rapidly from one country to the other through correspondence and personal visits.[22] The British participants in this intercourse considered themselves as members of an independent scientific and intellectual center. The supremacy of Paris was manifested in the fact that scientists and intellectuals from the other countries of

21 For England see Stimson, *op. cit.*, pp. 140 and 212–213. See also Annan, "The Intellectual Aristocracy," in J. H. Plumb (ed.), *Studies in Social History*, 1955, pp. 243–287, which deals mainly with the nineteenth century but shows that the situation was probably not too dissimilar in the eighteenth century. For France, see René Taton (ed.), *Enseignement et diffusion des sciences en France au XVIIIᵉ siècle* (Paris: Hermann, 1964), Sixième partie, "Cabinets scientifiques et observatoires," (articles by Jean Torlais, Charles Bedel, Roger Hahn, and Yves Laissus), pp. 617–712, which gives a good idea of how and by whom research was performed. An attempt by Yaacob Nahon at the Department of Sociology of the Hebrew University in Jerusalem to trace the social origins of forty-five scientists listed in Maurice Daumas (ed.) *Histoire des science*, Paris, *Encyclopédie de la pléiade* showed that at least 60 percent of those whose origins could be identified came from families with independent means. In any case, there were in France a few possibilities of careers in science teaching and research that did not exist in England. It also seems that relatively more French than British scientists came from families of doctors, lawyers, engineers, and civil servants. In England, on the other hand, many came from clerical families which, of course, did not exist in France.

22 Preserved Smith, *A History of Modern Culture*, Vol. II, *The Enlightenment, 1687–1776* (New York: Collier Books, 1962), pp. 331–333, 339–347.

84

the Continent treated France rather than Britain as their center and model. Their scientists and scholars went to Paris rather than to London for further study, and French was adopted as the *lingua franca* of intellectuals in Europe.

But England, although somewhat inferior to France in pure science, retained its intellectual and scientific independence. As a society where the scientific approach was deeply and broadly rooted in politics, economy, and social philosophy, it never became dependent on the French model, as did those societies whose institutions were hostile to science.

The Diffusion of Scientific Interest in Europe

We turn now to the diffusion of scientific activity in other European countries. Perhaps in some of the smaller countries of north western Europe and in northern Italy and Switzerland there existed a similar social background for the growth of science as in England and France. But in the other political units of continental Europe such as Russia, Prussia, Austria, or Spain there were no large and important social groups interested in the institutionalization of science as a social value. We cannot speak of a scientistic movement in these countries in the same sense as in the West. To the extent that there was such a movement, it was a secondary phenomenon, a transplantation of foreign ideas and social roles that had only weak roots in these societies. Still, these "secondary" movements were all successful to some extent, since the intellectual circles favoring them consisted of important, powerful, and able people (others had scant opportunities to learn about the state of science and of society in the West). They were also successful because the possible implications of science for technology (including military technology) was an important argument with every ruler. For the latter reason, the most important element in the French example was the academy, which was a device for the institutional insulation of science. All the absolutistic rulers of Europe wanted science for themselves, but they were afraid of the social consequences it might have. Thus there arose academies and other institutions for the support of science (often manned by invited foreigners) all over Europe, while private philosophical circles parallel to the French and British ones were few, relatively unimportant, and often persecuted.[23]

Separation of the Scientistic Movement
from the Scientific Community

The emergence of these secondary centers further enhanced the institutional distinction between the scientistic movement and expert science which had begun in France. The reasons were similar in both cases: natural science could be adopted in these still traditionalistic societies

[23] *Ibid.*, pp. 121–133.

because of its neutrality from the point of view of religion and conservative class traditions, but a scientistic philosophy which propagated social change could not be adopted.

This background precipitated the separation of expert science, or the scientific community from the scientistic movement. In fact the attitude of scientists toward the movement had been ambivalent from the beginning. On the one hand, it was obvious to scientists that their support came from this movement. They also understood that the scientific method had important implications for philosophical thought in general. Its minimal implication was that it invalidated most of the accepted ways of proof by abstract speculation and/or recourse to traditional authority; its maximal implication was that all affairs of society had to be approached in the same experimental fashion as was natural science. Because of this approach, scientists were usually sympathetic to the scientistic philosophical and social movements.

On the other hand, however, one of the most important aspects of experimental science is its precision and specificity. Every variable has to be measured, because differences so minute that they cannot be grasped even by imagination may decide whether a theory is right or wrong. Also, investigations are not guided by criteria of general importance as conceived by philosophers, but strictly by what is relevant to and soluble by existing theories and methods. The great fight for the dignity of modern natural science in the seventeenth century was partly a fight for the dignity of the precise step-by-step and operational approach of the scientist. This approach was programmatically emphasized by the Royal Society in its early stages, and it was strictly adopted by the *Académie des sciences*. From this point of view the broad intellectual aims of the scientistic movement were inconsistent with the specificity of scientific inquiry and constituted a threat to its integrity and specificity.

Opportunism also played a part in the emphasis of the distinctness and value neutrality of science. Official support of science in France and elsewhere on the Continent came from absolutistic, conservative rulers. The insistence on the strict neutrality of science and its specificity which made it accessible only to experts was, therefore, a condition of the freedom of scientific inquiry and a safeguard from nepotism and other governmental interference.

This specificity and value neutrality, which are part of the definition of empirical science, helped to create an international scientific community in Europe. While the social conditions prevailing in Britain, and to some extent in France, were necessary conditions for the emergence of this community, once these conditions existed in these two leading countries, the specificity and neutrality of science made possible its institutional insulation and also its growth on the entire Continent. This diffusion of science in many different types of societies and cultures helped to reinforce even further the separate identity of the scientific community. Networks of close communication arose between scientists in Europe that increasingly excluded amateurs and the general philosophers.

The separation of amateurs of the scientistic movement from experts of the scientific community created a new situation toward the end of the eighteenth century. On the one hand, elements of scientific professionalism began to appear; on the other hand, as a result of their academic privileges, scientists became to some extent a part of the privileged classes of the traditionalist society. The implications of this development will be explored in the next two chapters.

the rise and decline of the french scientific center in a regime of centralized liberalism

six

The shift of the scientific center from England to France that occurred during the second half of the eighteenth century did not establish a very pronounced French superiority. France was the center of the scientific world, but for England it was only a slightly more successful competitor, and no more than a senior partner in a common intellectual enterprise. During the first three decades of the nineteenth century, however, French scientific supremacy became much more unequivocally established.[1] In spite of the

[1] In the 26 five-year periods between 1771 and 1900, Britain made more discoveries in heat, light, magnetism and electricity than France or Germany in eight. Germany made most discoveries in these fields in 11, France led the other two countries in six and Britain and France were tied for leadership in one of the five-year periods. Of the eight half-decades of British superiority, seven of them fall between 1771 and 1810; of the 11 half-decades of German superiority, 10 fall between 1851 and 1900 when Germany led in every five-year period. France's periods of supremacy lay between 1815 and 1830, with a renewal of supremacy between 1841 and 1850. See Rainoff, T. J., "Wave-like Fluctuations of Creative Productivity in the Development of West European Physics in the Eighteenth and Nineteenth Centuries," *Isis*, XII, 2 (May, 1929), pp. 311–313, tables 4–6. In physiology, between 1800 and 1924, France led in a number of original contributions in three out of the five half-decades until 1824. After that Germany led in every half-yearly period until the end of the period. A similar calculation of numbers of discoveries in medical sciences by decades between 1800 and 1926 shows France leading from 1800 to 1829, and Germany taking the lead thereafter in every decade until 1910 when the lead passed to the United States. See Zloczower, A., "Analysis of the Social Conditions of Scientific Productivity in 19th Century Germany" (unpublished M.A. dissertation, Hebrew University of Jerusalem), based on Rothschuh, K. E., *Entwicklungsgeschichte physiologischer Probleme in Tabellenform* (Munich: Urban & Schwarzenberg, 1952), and Ben-David, J., "Scientific Productivity and Academic Organisation," American Sociological Review, XXV, 6 (December, 1960), p. 830 and tables 1, 2, 5 in the Appendix.

88

brilliance of some British scientists such as Dalton, Davy, Faraday, and Young, neither in Britain nor anywhere else were there first-rate scientists covering all the then existing fields of science. Only in France, or more precisely in Paris, were all fields of science pursued at an advanced level.[2]

This systematic coverage of all the sciences in a single center has been interpreted as the first instance of organized, professional science in contrast with the amateur pattern of the seventeenth and eighteenth centuries. This view would seem to be supported by the existence of a few institutions of scientific higher education in France which were more advanced than those in other countries. But it is difficult to make this interpretation consistent with the fact that French scientific leadership came to an end about 1830 to 1840. Had this really been the first instance of organized professional science, then the scientific supremacy of France should have lasted longer than three decades. The professionalization of science yields its best results in the second or third generation.

The great upsurge of French science following the Revolution was only indirectly related to the new institutions of higher education established between 1794 and 1800, and those institutions did not constitute a beginning of organized professional science. They were rather the culmination of eighteenth century patterns of scientific work. Furthermore, the upsurge was due to the re-emergence and reinforcement under Napoleon and the Restoration of the same constellation of social forces which furthered the growth of science during the last decades of the *ancien régime,* and which was temporarily disrupted under the Revolution. This interpretation is consistent with the duration of the upsurge and the paradoxical onset of the decline in the 1830s when a liberal regime finally made the "institutionalization" of scientific values in France possible.

The first part of this chapter is an attempt to substantiate the foregoing interpretation. The second part attempts to explore the structure and the working of French science during the rest of the nineteenth and twentieth centuries. The main purpose of this will be to see why, in spite of its numerous parallels with the British system, French science has been relatively ineffective in responding to the challenge of organized scientific research which emerged in Germany about the middle of the nineteenth century and was subsequently further developed in the United States.

The Importance of Scientism in the
Advancement of Science

The pressure during the Revolution for the establishment of new scientific-educational institutions and regular careers for scientists, scholars, and philosophers was primarily a product of demands by scientistic philosophers and other intellectuals rather than of demands by expert scientists. The way this occurred was determined by the ambiguous relationship between

[2] See Maurice Crosland, *The Society of Arcueil: A View of French Science at the Time of Napoleon* (London: Heinemann, 1967), pp. 429–467.

the scientistic movement and expert science in its academic form, which was discussed in the previous chapter.

The scientistic movement in French intellectual opinion consisted from its very onset of persons with practical interests in politics and economics. Their principal aim in using science as a model in political and economic affairs was to provide objective, "scientific" proof of the necessity for changes which they desired, and which they could not or would not support by traditional arguments. They were often careless and superficial in their thinking. There was a great deal of confusion about the meaning of scientific laws when applied to human action, and much confusion between statements of fact and judgments of value. This confusion persisted throughout the eighteenth and nineteenth centuries in much of the philosophical thought about man and society.[3]

Nonetheless, until the second half of the eighteenth century the philosophers did not raise the question as to whether there were methods other than empirical science to arrive at truth through efforts of human intellect. Explicitly or implicitly, they accepted Newtonian natural science and Baconian intellectual strategy as the only available method, short of revelation, of attaining significant and objectively valid knowledge. Their purpose was to explore what this method and knowledge consisted of and to apply the conclusions to morals, politics, and economics.

In the eighteenth century Britain was the only large country where people could propagate change and reform without the danger of persecution. Furthermore, the intellectuals in Britain were an integral part of what can be described as the upper middle class. They were often wealthy and well-connected, and many earned their incomes from ecclesiastical or governmental positions, or independent professional work. They were not politicians themselves, but they usually had direct access to political leaders and often acted as their advisers. It is not surprising that they had first-hand and practical knowledge of politics, economics, and legislation and that their views were rarely revolutionary or utopian. The way they tried to apply science to the practical problems of society was not dissimilar to the way the great inventors of the age went about the application of the scientific approach to the building of machinery or the treatment of illness. They were well aware of the complexity as well as the specificity of the problems with which they dealt and did not try to deduce suggestions for social reform from first principles. Even a man like Bentham, who was inclined by temperament to reason from first principles was driven to a great deal of practical social gadgeteering.[4]

Germany was almost at the other end of the scale, as it was a country (or rather an area comprising several countries) where change was legitimate only if initiated by the ruler and where intellectuals (except some foreigners) had

---

[3] Charles C. Gillispie, *The Edge of Objectivity: An Essay in the History of Scientific Ideas* (Princeton: Princeton University Press, 1960), pp. 151–157.
[4] Shirley Letwin, *The Pursuit of Certainty* (Cambridge: Cambridge University Press, 1965), pp. 176–188.

rise and decline of the french scientific center under centralized liberalism

no access to policy making. Hence there was little incentive to treat political and economic questions in the same manner as in Britain and France. France, finally, was somewhere in between. The place of the intellectual in French society was similar to that in English society. The outstanding ones among them were members of the upper middle class and had excellent contacts with the ruling circles. At the same time, however, France was in many ways ruled in an even more traditional manner than Prussia and other German lands. Religious pluralism was not officially tolerated, invidious distinctions of status and rank were officially bolstered, and attempts at social reforms had to stop short at sacrosanct traditional prerogatives.[5]

Another aspect of the situation was the degree of social consensus concerning the legitimacy of change. In Britain, social heterogeneity and social change were generally accepted, and even those who considered these things as evils were resigned to living with them. In Germany, change and heterogeneity were only acceptable to a few, although these few had often disproportionate influence in ruling circles. In France, however, society was much more evenly divided between those who favored and those who opposed change, and the balance of power between the two factions was much more delicate than anywhere else. Since the church and the official educational establishment in general (with very few exceptions, such as the Collège de France and in part, the academies) were monopolized by the traditional sector, the clash between progressives and traditionals was particularly violent in matters of education and religion.

The effect which this situation had on political thought in France has been described by de Tocqueville, who believed that if ideas about society were not put to the test, there was no way to judge their effects. The ideas, therefore, became increasingly abstract and doctrinal.[6] This was his interpretation of the situation in France in the period before the Revolution. Furthermore, since the intellectuals realized, or at least believed, that they could not change anything, the purpose of their writings was to make a striking intellectual impression and to stir opinion.

This tendency was further reinforced and channeled in a new direction by the invidious distinction conferred upon natural scientists by the same governments that persecuted and despised other intellectuals, and the supporters of the scientistic movements among merchants, technologists, and upper grade artisans. As a result, there was ambivalence toward science even among the groups which had originally been its most ardent supporters. On the one hand, these groups were still interested in the establishment of a freer society; change and improvement through social reform was still the aim of these philos-

[5] Preserved Smith, A History of Modern Culture, Vol. II, The Enlightenment, 1687–1776 (New York: Collier Books, 1962), pp. 483–490. Smith shows that the Prussian press was considerably free in religious matters and even in satirizing the regime. But criticism of the actual conduct of the government was not allowed, and the views of the intellectuals carried no weight with the government.

[6] Alexis de Tocqueville, L'Ancien Régime (Oxford: Basil Blackwell, 1937), pp. 147–157.

ophers as well as of their scientistically minded public. Science that seemed integral to these reforms continued to be an important symbol for them. On the other hand, they wanted a type of science in which they could participate and which would be relevant to their aspirations.[7]

This situation led to a questioning of the very validity of Newtonian science as a model for the logic of inquiry. Attention turned to contents and methods of cognition which were not assimilable to science as it was conceived at that time. For example, Diderot chose to play up chemistry and biology against mathematical physics as the model for sciences, and Rousseau pointed out the inadequacy of science for the description of the moral experience of man and suggested a new intuitive conception of nature as the valid way to true understanding.

The questions raised by Diderot and Rousseau were all valid and were as implicit in the state of natural and social science as those raised by Locke and Hume. In this sense these questions emerged out of immanent intellectual developments. Indeed the problem Diderot raised, the handling of complex structures, became the central achievement of organic chemistry, biology, and electromagnetism in the nineteenth century. The inadequacy of eighteenth century science in explaining problems successfully handled by technology was epitomized in the discovery of the steam engine.[8]

The same is true of the problems raised by Rousseau. The basis of validity of moral values in secular societies and the place of creative intuition in the scheme of scientific method have been basic questions to which the social sciences have returned time and again since then. Speculations about the moral consensus of modern societies eventually led to the foundation of modern sociology by Weber and Durkheim.[9] The place of intuition and metaphysics in scientific discovery remains an intensely debated question among philosophers even today.[10]

The long-term effects of this development were very significant. Parallel with an empiricist social science which tended to select problems amenable to empirical investigation and to leave the basic metaphysical questions for ceremonial occasions (as happened in Britain), there emerged in France a tradition of raising basic philosophical questions with little regard for their practical consequences or empirically demonstrable solutions. The raising of basic issues, however justified they are in principle, only contributes to knowledge on rare occasions and is usually avoided by "normal science." [11] The British

[7] Gillispie, *op. cit.*, pp. 178–201.

[8] *Ibid.*, pp. 173, 184–192.

[9] Talcott Parsons, *The Structure of Social Action* (New York: McGraw-Hill, 1937), pp. 307–324; Raymond Aron, *Main Currents in Sociological Thought*, Vol. I (New York: Basic Books, 1965), pp. 89–91, 198–202, and Vol. II (New York: Basic Books, 1967), pp. 11–23.

[10] Thomas S. Kuhn, *The Structure of Scientific Revolutions* (Chicago University Press, 1963), pp. 84–90.

[11] *Ibid.*, pp. 35–36, 76–79.

92

social philosophers of the eighteenth century behaved in this respect like "normal scientists." Even when, like Hume, they raised fundamental problems, these were never carried to the extreme where they became abstract doctrines which questioned the validity of all moral order and rational search for social reform.[12] In France speculation was carried to just such extremes. Even if this was not the intention of the philosophers, their ideas could easily be used in political or ideological rhetoric for attacking the very foundation of political and moral order.[13]

What is of interest here is not the validity of these questions, but the effects of raising them. Implicit in this development was the rise of new intellectual movements which were secular and ascientific (or potentially even antiscientific), and accepted neither the discipline of religious orthodoxy nor that of scientific method.

These, however, were not the immediate consequences. Diderot and Rousseau belonged to the Enlightenment, and the short-term effect of their ideas was to make men more optimistic about their own intellectual and moral capacities. This implied a loosening of discipline and the sense of responsibility in philosophical and even in scientific thought, but not, however, a rejection of natural science as the model of intellectual inquiry nor the rejection of the political and educational ideals of the scientistic movement. The combination of the loosening of intellectual discipline with adherence to the goals of scientism turned the latter into a popular movement.[14] Thus the frustration of the political and social goals of the scientistic movement led to the taking up of philosophical themes for the sake of literary accomplishment rather than for the purpose of providing solutions for practical problems. This in turn popularized the scientistic movement and generated a great intellectual ferment, as manifested in the foundation all over France of local academies, literary circles, and clubs eagerly discussing science, social and economic problems, and philosophy.

The popularization of scientific and philosophical interest reversed the trend toward greater separation between expert science and the scientistic movement. The popular discussion of science insisted that science be socially, technologically, and politically "relevant." Although this demand for "relevance" contained seeds of potential antiscientism and more than potential scientific quackery (such as the famous case of Marat), it also implied admiration for science and a willingness to support it and use it as widely as possible.

The intellectuals became completely alienated from the official educational establishment, especially from the Sorbonne, controlled by the church (and, in the appropriate faculties, by the medical and legal guilds). As a rule they were

[12] Elie Halévy, The Growth of Philosophic Radicalism (Boston: The Beacon Press, 1955), pp. 11–13; as has been pointed out, this was true even of the most doctrinaire of the British philosophers, Bentham.

[13] Lester G. Crocker, An Age of Crisis: Man and the World in 18th Century French Thought (Baltimore: Johns Hopkins Press, 1959), pp. 9–106, 461–473.

[14] D. Mornet, Les Origines Intellectuelles de la Révolution Française 1715–1787 (Paris: Armand Colin, 1934), pp. 35–95, 125–127.

93

also opposed to the Roman Catholic Church. Those participating in the new current of intellectual activity felt persecuted although they were not actually suppressed by the official intellectual establishment. They considered the powers and the official privileges of the institutions and the intellectuals who staffed them as completely illegitimate.[15]

Revolutionary and Napoleonic Reform
of Intellectual Institutions

Following the devastation caused by the Reign of Terror, a new educational and scientific structure was created, and the secular intellectuals acceded to the intellectual monopolies previously exercised by the clergy. It was this, and not the internal requirements of science, which led to the emergence of new educational organizations and government offices providing careers for secular intellectuals, including scientists. The scientistic outlook of the intellectual movement and its high regard for science set the structure of the system. At the top of the system were a number of *grandes écoles* (including some established under the *ancien régime*), designed to train personnel for government service and higher (including the upper levels of secondary) education. The most famous of these were: the *Ecole polytechnique*, for the training of civil and military engineers; the *Ecole normale*, designed to create a new body of professors for the upper levels of education which after a number of changes came to include the *lycées*, the *collèges* and the faculties (the universities in the accustomed sense were abolished in 1793 and were only nominally reestablished in 1896, so that every faculty constituted a separate institution); and the *Ecole de santé*, later *Ecole de médecine*.

These educational establishments had, by the standards of the times, excellent laboratory facilities and were complemented by the *Institut*, the *Musée d'histoire naturelle* and the *Observatoire* which were purely research establishments, or in the case of the *Institut* research and honorific establishments.[16] But they did not represent any new conceptions either in teaching or in the organization of research. Specialized schools for advanced training for the various professions were favored by the "enlightened" absolutist regimes and had already existed before the Revolution (*e.g.*, the *Ecole de ponts et chaussées*). Even elite institutions of a purely scientific nature such as the *Collège de France* and the *Musée d'histoire naturelle* assumed their dominant character before the Revolution.[17] The *Collège de France* became an even more distinguished institution, where all fields of science and scholarship were taught at the most advanced

[15] *Ibid.*, pp. 129–134, 150, 177, 270–281.
[16] Charles Newman, *The Evolution of Medical Education in the Nineteenth Century* (London: Oxford University Press, 1957), p. 48, and Maurice Crosland, *op. cit.*, pp. 190–231.
[17] Ernest Lavisse, *Histoire de France Illustrée*, Vol. IX (Paris: Librairie Hachette, 1929), pp. 301–304, and René Taton (ed.), *A General History of the Sciences*, Vol. III, *Science in the 19th Century* (London: Thames and Hudson, 1964), pp. 259–440, 511–615.

rise and decline of the french scientific center under centralized liberalism

scientific level in a spirit of academic freedom; the spirit of free inquiry was also introduced into some of the new specialized schools, such as the *Ecole polytechnique*, which were ostensibly for the training of professional practitioners.[18]

The Position of Research in the New Institutional System

Only the "central schools" of sciences and letters, designed to replace the church-controlled *collèges* which had been abolished in 1793, represented a new educational experiment qualitatively different from anything which had existed before. Although intended to be higher secondary institutions, they were in many respects the first attempt to establish a modern university.[19] Had these schools been maintained, they might have led to the emergence of regular careers in research and to modes of organized research such as eventually developed in Germany. But the experiment was soon abandoned. The specialized schools, although now more numerous and of a higher standard than before, perpetuated the eighteenth century patterns of the scientific role and scientific work. Their teachers were supposed either to train students preparing to take special examinations and enter particular careers, or to lecture freely to an undifferentiated audience. Neither of these activities entailed the transfer of the locus of research to the institutions in question or the involvement of the student in the teacher's research. Although the majority of scientists became teachers, research continued to be a private activity, as it had been before the Revolution, when scientists earned their living from a variety of sources. Teaching was a partial sinecure which provided an opportunity to engage in research; it was not regarded as having anything more to do with research than that. Other partial sinecures, such as certain civil service appointments, were acceptable alternatives.[20] As far as research was concerned, the amateur pattern still prevailed.

The isolation of research from teaching continued because there was no intellectual or economic incentive to overcome it. As has been pointed out, had the central schools been maintained there might have arisen a demand for a combination of the teaching and research roles, but as things were there were good arguments against such a combination.

In seventeenth- and eighteenth-century Britain and France, humanistic studies were not included in the conception of science. The question as to the extent to which the scientific method should be applied to humanistic studies became an important issue in these countries only during the nineteenth century

[18] Gillispie, *op. cit.*, pp. 176–178.

[19] Louis Liard, *L'enseignement supérieur*, Vol. II (Paris: Armand Colin, 1894), pp. 1–18, and Georges Lefebvre, *The French Revolution: From 1793 to 1799* (London: Routledge & Kegan Paul, 1964), pp. 290–292.

[20] Crosland, *op. cit.*, pp. 1–5, 70, 151–179, about the employment of French scientists. The interpretation that this was a continuation of the amateur pattern is not Crosland's, but mine.

under the influence of German scholarship.[21] This is not to say that French scholars in these fields in the early nineteenth century were not as outstanding as any. In some fields, such as Oriental studies, Paris was indeed the world center.[22] But it was accepted that the humanistic subjects had important aesthetic and moral aspects which distinguished them from science. The importance of these aspects to education was never denied. On the other hand, natural science and the scientistic social sciences were supposed to have practical applications for technology, economy, and government which the humanistic subjects did not possess. Thus the overlap between the function of the humanities and the newer scientific fields was only partial, and there was no awareness that all these different fields might be optimally pursued within a single organization by persons who adopted similar methods of investigation and instruction and considered themselves to be members of the same profession.

This explains why the capture of the uppermost reaches of the educational system and parts of the civil service by scientists (and prior to the Napoleonic empire, by scientistic philosophers), did not lead to a complete "scientization" of the educational system such as occurred eventually in Germany. In this latter country, even the humanities came to be taught both at the higher and at the secondary level in a scientific way based on systematic philology.

While every scientist and philosopher in the revolutionary period was convinced that education needed thoroughgoing reform, and especially a great infusion of scientific content, there was no sense that there was anything wrong with scientific research. French scientists did better than any others within the existing individualistic pattern of private laboratories. There was no demand or wish to change this pattern and shift the locus of research to the educational establishments. Thus, while the revolutionary period can justly be seen as the beginning of scientific educational policy, it was not the beginning of a deliberate science policy.

The distinction between education (imbued with science) and research was also apparent from the attitude toward academic freedom. The apparent lack of this freedom in the nineteeth-century French system has often been noted. Nonetheless, French research workers were as aware as others of the importance of scientific freedom, and there is no evidence of any interference with the freedom of research. Education, however, was a different matter. French scientists and scientistic philosophers were insistent on removing church control from the school system, but they were not interested in eliminating the direct control of the state—which had its own nonscientific interest in the creation of a loyal citizenry—or in dictating to educationists in nonscientific fields how to go about their jobs. In fact, they saw strict state control of education as a necessary safeguard against the resurgence of church control.

[21] Terry N. Clark, *Institutionalization of Innovations in Higher Education: Social Research in France, 1850–1914* (unpublished Ph.D. dissertation submitted to the faculty of political science, Columbia University, 1966), pp. 319–321.
[22] Liard, *op. cit.*, pp. 172–173.

rise and decline of the french scientific center under centralized liberalism

There was no reason for the scientists to object to the state control of education, since they themselves were prominent in the civil service and especially in educational administration.[23] The educational reformers looked forward to the creation of a society where science and technology were to play a leading role. They envisaged a state where economic production and social welfare would attain new heights as a result of the contributions to research and discovery of a brilliant corps of scientific and technological leaders and the work of a prosperous and patriotic citizenry.[24] Teaching science was, therefore, only one of the many tasks assigned to scientists and was not to be their exclusive domain. The freedom of the scientists was ensured by the privacy of research. All that was needed for this was some private means and a few public facilities for competent scientists. This need was adequately taken care of by the eighteenth-century pattern, which was greatly augmented in scale by the new opportunities.

Why French Science Flourished in the First Third of the Century

The Napoleonic policies further reinforced the eighteenth-century pattern. The discontinuation of the central schools and the reestablishment of traditional syllabi in primary and secondary and in part of higher education abolished the only potential source of change. But none of the Napoleonic policies caused any harm to expert science, nor did they reverse the trend of increasingly using the services of scientists for a variety of functions. The spirit of high-grade science prevailed in some of the *grandes écoles* and in a few faculties.[25] The links of the leading scientists with the political elite, which began on a very small scale during the last decades of the *ancien régime*, were considerably extended. Scientists as a class, and not just a few privileged ones, became part of the official elite during the last years of the Revolution, and they maintained this status under Napoleon. Berthollet, Cuvier, Laplace and others were given important positions in government and/or were trusted advisers to the Emperor. The increasing autocracy of the Empire and the reactionary policies of the Restoration perhaps reduced the actual influence of the scientists, but these developments did not reduce their potential influence since they remained members of the elite.[26]

This opening-up of new opportunities for scientists is evident from a survey of the occupations of scientists before and after the Revolution. Before 1789, the majority of scientists were wealthy men (noblemen, physicians, etc.) who financed their scientific work from their own resources. Even Lavoisier,

[23] For the importance of scientists in the late revolutionary period and under Napoleon see Crosland, *op. cit.*, pp. 1–5, 70, 151–179.
[24] John Theodore Merz, *A History of European Thought in the Nineteenth Century*, Vol. I (New York: Dover Publications, 1965), pp. 110–111, 149–156.
[25] Liard, *op. cit.*, pp. 57–124.
[26] Crosland, *op. cit.*, pp. 4–5, 20–26, 42.

rise and decline of the french scientific center under centralized liberalism

who was as close to being a career scientist as anyone could be at that time, had to maintain himself as a tax farmer and could devote only one whole day a week to his scientific work; the other days were divided between research and business.[27]

Table 6–1

Types of Career of French Scientists Born During the Eighteenth Century

Born	Traditional[a]	Modern[b]	Traditional to Modern[c]	Unknown
1745 or before	31	8	10	0
1746–1755	14	3	9	1
1756–1769	5	5	8	1
1770–1789	6	34	2	2

[a] Traditional: Priests, lawyers, physicians, industrialists, engineers, proprietors, army officers, and civil servants whose work was unconnected with education.
[b] Modern: Teachers, research workers, and civil servants connected with education.
[c] Traditional to modern: Those who switched from one type of career to another.

SOURCE: The table is based on a cumulative list of scientists collected from different books and biographical collections with a view to obtaining as complete a list as possible.

After 1796 scientists in France usually had a position either in higher education or in the educational civil service (or occasionally in some other civil service position, presumably granted to them for their scientific achievements).

Thus the reason for the dissatisfaction with the educational and general intellectual monopolies of the clergy disappeared. This fact was not changed by the Napoleonic reaction against the revolutionary reforms. Scientists certainly had no reason to feel that they were prevented by traditional status privileges or ecclesiastical monopolies from using their talents and reaping the social benefits due to them.[28]

Even the moral problem of acquiescence in the new situation was less severe than it may appear. The situation which inspired the educational reforms of the late revolutionary period had passed. In a closed class society where power, honor, and economic means were allocated to organized estates, the immediate goal of all "modern" intellectuals was the replacement of the existing intellectual estate (the church and the university corporations) by themselves. With the abolition of the estates, however, the whole perspective changed. At the very point when the scientistic movement obtained the educational monopoly, this monopoly lost its value as a means of ensuring the dignity and resources for scientists and philosophers. Once the whole society was open to them, education became a much less important issue.

[27] Gillispie, *op. cit.*, p. 215.
[28] Lefebvre, *op. cit.*, p. 305.

rise and decline of the french scientific center under centralized liberalism

Finally, the scientists were not alone in abandoning the educational ideas of the Revolution. The great intellectual ferment embracing broad classes of society preceding the Revolution had abated. The autocratic educational policies of Napoleon were probably not much different from what they would have been under a system of laissez faire, except that in the latter case there might have been more experimentation and more variety. There was probably as little popular sympathy for the continuation of the scientistic trend in education as for the continuation of the general revolutionary upheavals.[29]

Similarly, there was little enthusiasm for the continuation of the educational utopia of Condorcet which would have created opportunities for higher education at the most advanced level for everyone capable of benefiting from it intellectually. It was not easy to find competent teachers and students for the central schools throughout the country. And there was no interest in force-feeding the educational system so that it might serve as a mechanism for the equalization of social status. Having attained the abolition of legally defined estates, few Frenchmen were interested in meddling further with the class system.[30]

The end result of the revolutionary and Napoleonic reforms was, therefore, an enhancement of eighteenth-century patterns and conceptions of scientific work. The peak of the organizational system consisted of the *Institut* and the *grandes écoles* which were all prerevolutionary structures. Furthermore, the new *grandes écoles* no longer had to contend with the privileged, nonscientific universities. Those had been abolished and the faculties which came in their stead were less privileged than the *grandes écoles*. Finally, some of the faculties also taught science, and a limited amount of science teaching was also introduced into secondary education.[31]

The great flowering of French science between the years 1800 and 1830 was, therefore, not the result of any new ideas or practices about scientific training and research, or about the uses of science. It was rather the result of increased support for science, and probably increased enthusiasm for it in the eighteenth-century manner. These were generated by the same conditions which had existed before the Revolution. The excesses and the upheavals of the latter created a reaction against political and educational reforms and ideological pre-occupations. At the same time, however, the classes supporting the scientistic movement became much stronger. Successive French governments, even if reactionary, had to reckon with them and conciliate them. This was the same constellation as that which prevailed under the Restoration in Great Britain and during the last decades of the *ancien régime*, but the balance was further tilted in favor of science.

[29] Paul Gerbod, *La Condition Universitaire en France au XIXᵉ Siècle* (Paris: Presses Universitaires, 1965), pp. 78–81.

[30] Lefebvre, *op. cit.*, pp. 291–309.

[31] For the state of higher education following the Napoleonic reforms see Liard, *op. cit.*, pp. 119–124.

This spirit was reflected in the way research was supported. Some of the *grandes écoles* were given lavish facilities. This support, however, was not done with a view to creating public facilities for the systematic training of future research workers, as the purpose of training was practical, and did not include the preparation of theses. It was rather a public gesture in favor of science. The evidence for this is that there was no policy for keeping the facilities up to date or for developing them in accordance with the changing requirements of science and the numbers of students who were to be trained.[32]

Nevertheless, in a few cases, the new facilities were effectively used for the purpose of training relatively large numbers of students, and in any case they provided increased opportunities for research as well as for the acquisition of scientific knowledge, especially as the motivation to study science and excel in it was high. Hence the opportunities were taken advantage of by the older scientists who survived the Revolution as well as by the generation which grew up during the Revolution. The meeting of these two generations was, therefore, an extremely fruitful one, and the transition between them resulted in a great rise in the level of scientific activity.[33]

Stagnation and Decline after 1830

This period of activity was followed by stagnation and subsequent relative decline of scientific activity in France during the 1830s and the 1840s. After the Napoleonic period the situation of science in the class structure of French society became similar to that prevailing in England. Science was now "institutionalized" in the sense that scientists, and after the brief interval of suppression scientistic intellectuals as well, could aspire to all the honors and influence they might have wished for. It was possible to use science and to apply it as widely as possible, and any success in this respect was greeted by social approval.

Therefore, once the opportunities provided by the abortive reforms of the 1790s were exploited, there was no more drive to change the educational and scientific systems. When the Napoleonic era and the Restoration came to an end, there was an opportunity to resume the work of the Revolution in educational reform. But the attempts to do this were thwarted by the priority given to political considerations over scientific and educational interests.[34] As in England after the Glorious Revolution, in France the "institutionalization" of

---

[32] *Ibid.*, pp. 209–218.

[33] Crosland, *op. cit.*, pp. 97–146 for life histories of scientists connected with the Society of Arcueil which illustrates this meeting between the two generations.

[34] The idea of updating the faculties and making them autonomous was put forward by Guizot and Cousin, but it was dropped for fear that any loosening of state control over the conferment of academic grades might be used by the church for the strengthening of its own system of education. This apprehension was enough to thwart the reforms since there was no noticeable demand for a higher level of scientific and scholarly education. See Liard, *op. cit.*, pp. 179–199, 215–217.

100

science led to a relative decline of scientific enthusiasm. Once the opportunities opened up by the revolutionary changes were exploited, there was a deflection of interest to social reform, social philosophy (Fourier, Saint-Simon, Comte), and technological activity.

Thus, by the 1830s, science had lost the symbolic glamor which it had possessed in the eighteenth century and which had been further enhanced during the first decades of the nineteenth century. France was a society which offered many other attractive opportunities. A young man whose talents allowed him to choose between science and more practical interests in 1780 would probably have tried his luck at science first.[35] By 1840, he would probably have been drawn more to practical politics, business, industry, or perhaps creative writing.[36] All these allowed him as much freedom as science, and an equal or superior income.

Thus scientific growth settled down to a pattern similar to the English one. Creative scientists were not produced or trained by any particular part of the educational or scientific system. They were either individuals with a strong sense of personal calling and of exceptional genius, or members of families with a strong tradition of scientific interest and perhaps hereditary talent, who sought their teachers at the Sorbonne, the *Collège de France*, the *Ecole normale*, or wherever they happened to be.

Accordingly, the development of French science up to the decade between 1830 and 1840 can be explained as a function of the degree of institutionalization of scientific values. The support for it was generated by a belief in a pragmatic and "progressive" social order. Where this belief was not shared by politically and economically important groups of the population, there was, on the whole, little scientific activity. Where it was so supported, the volume of scientific activity varied according to the degree of the realization of the general social aspirations of the groups supporting science (i.e., the scientistic movement). In the British as well as the French case, the peak of the support for science (including, apparently, personal motivation, as well as the establishment of official scientific institutions), was reached during the lull following violent revolutions, just before the liberal reforms demanded by the scientistic

---

[35] Marat was a good example of the attraction which science held for an ambitious young man whose talents were journalistic and political prior to the Revolution. See Louis R. Gottschalk, *Jean-Paul Marat: A Study in Radicalism* (New York: Greenberg), pp. 8–31.

[36] See Liard, *op. cit.*, pp. 211–222, for an account of the relative lack of interest in science and scholarship and the abandonment of academic careers for careers in politics during the 1840s. The small attraction that science held after the Napoleonic reforms which lasted until 1880 can be seen from the very small number of diplomas in science granted by French faculties. Only in the decade from 1861 to 1870 did the yearly average surpass the 100 mark; prior to that the average was considerably less. Thus the total output of science graduates from all the French faculties was probably less during this period than the output of the one engineering school, *the Ecole centrale des arts et manufactures*, which graduated some 3,000 engineers between 1832 and 1870, an average of approximately 75 a year. See Antoine Prost, *L'Enseignement en France 1800–1967* (Paris: Armand Colin, 1968), pp. 243, 302. For the widespread initiative—much of it private—in technical and technological education during the first half of the nineteenth century in France, see F. B. Artz, *The Development of Higher Technical Education in France* (Cambridge, Mass.: The MIT Press, 1966), pp. 212–268.

rise and decline of the french scientific center under centralized liberalism

movements were firmly established. During these transitional periods revulsion from the violence and anarchy of revolutions temporarily halted the push for the realization of the broader aims of the movement, as well as the preoccupation of many of its adherents with philosophical (or in the English case, theological) and educational problems, thus centering attention on science. The flattening out of the growth in scientific enthusiasm began in both cases with the relatively peaceful establishment of liberal regimes and the dispersion of intellectual interest in political, economic, and technological concerns. Now that change was possible and legitimate in every sphere of life, there was no more reason to concentrate innovative talent and interest on science alone.

This explains the apparent paradox that the return to liberalism in 1830 did not bring in its wake a return to the scientific enthusiasm and educational reforms of the revolutionary period. But it does not explain the subsequent course of French scientific growth. Toward the middle of the century the conditions of this growth changed. It ceased to be exclusively determined by the preferences of the intellectual community and its supporters; to an increasing degree it came to depend on the organization of higher education and research. Such conditions emerged first in Germany and presented a challenge to the older scientific countries, such as Britain and France. It is difficult to understand the relative inability of the French system to respond to this challenge, which became increasingly evident in the 1840s. In France, as in England, science was accepted by and large as an intrinsically worthwhile pursuit as well as a generalized tool of social amelioration. The resources at its disposal early in the century were superior to those existing in England. How then can we explain the fact that English science, when faced with growing German and later American superiority, could reform itself promptly and effectively, and as a result enter a period of steady growth, while the French response occurred later and did not lead to uninterrupted growth?

This problem cannot be solved by the scheme of analysis which has been applied so far. The explanation is not to be sought in the interests of wider social groups in science or in a scientistic philosophy (which were similar in France and England), but in the peculiar characteristics of French scientific organization.

The outstanding feature of this organization and of French bureaucracy in general was centralization. Whether as a result of age-long absolutistic traditions, or because of the basic rift in French society between those who welcomed the Revolution and those who never accepted its legitimacy, the French civil service never renounced its prerogatives of control over every aspect of social life.

This control had a variety of debilitating effects. In order to maintain it, the government preferred to establish schools and institutions with very specific purposes. Science, however, was rapidly changing, so that what was an adequate organization in 1820 was likely to be out of date 20 years later. To keep pace with developments, scientific organization should have been constantly adapted

rise and decline of the french scientific center under centralized liberalism to new situations. But it was difficult to change organizations that had narrowly defined purposes without the use of coercion. Furthermore, the more centralized a system, the greater the likelihood that even relatively minor changes in the existing state of affairs would have unexpected political or administrative repercussions. Under such circumstances, it was preferable to create new institutions rather than to try to change existing ones. In order not to hurt vested interests and also for reasons of bureaucratic convenience, these again had to be "special purpose" institutions.

The Ecole Pratique des Hautes Etudes

The way in which this system limited the effect of even the most imaginative innovations can be seen from the case of the *Ecole pratique des hautes études* established in 1868. This school can be considered as the first experiment in postgraduate training. Its purpose was to organize courses, seminars, and laboratory instruction conducted in a completely free manner by the outstanding research workers in Paris, irrespective of their affiliation to one of the faculties or *grandes écoles*. Through this device all the outstanding but scattered and fragmented scientific talents were pooled for the purpose of advanced training for research. The conception was more advanced than anything that existed in Germany or elsewhere at that time, since nowhere else did there exist a specialized scheme for the training of research workers.

There is little doubt that this institution made an immense contribution to the training of scientists and scholars. Having been conceived from the very outset as a complement to existing institutions, however, its potentialities for development were greatly limited. At the time of its establishment, the *Ecole pratique* was denied the power to grant degrees. In the long run the lack of this power in the only institution designed to train research workers at an advanced level delayed the correction of one of the most anomalous features in the French academic career, namely, the requirement that aspirants to such a career should pass an examination (the *agrégation*) rather than prepare an advanced piece of research.

The long-term disadvantages arising from the absence of students properly "belonging" to the *Ecole pratique* was paralleled by the absence of a teaching staff exclusively or preponderantly identified with the school. This reduced the incentive for initiating changes and innovations in its structure. It also restricted opportunities for cooperation or even significant intellectual interchange between members of the teaching staff.

Inflexibility as a Result of Centralization

This institutional inflexibility was reinforced by assigning the *Ecole pratique* a special function to supplement those performed by other

103

institutions.[37] Had the situation been such as to allow competition among a number of institutions, then its example might have spread more widely in France. In the United States, and to some extent even in Britain, any successful innovation in higher education was bound to be imitated and reproduced in several institutions. The competition which thus arose stimulated further changes and innovations. The centralized French system, where each institution had a special and rigorously delimited function, produced exactly the opposite results. The success of a single institution made it unnecessary to duplicate a function already so well taken care of.[38] Thus academic anomalies were permitted and indeed forced to survive.

Of course, the avoidance of "unnecessary overlap" and "duplication," and the use as far as possible of existing resources and manpower are highly reasonable administrative principles. But their application resulted in France in the continuation of a vicious circle in which pioneering institutions established by enlightened administrations were too narrowly based and too rigidly fitted into the whole structure to be able to influence the system or to adapt themselves when the need for adaptation eventually arose. Even where they were able to maintain high scientific standards, they were incapable of the initiative and rapid expansion which characterized similar institutions elsewhere.

One way of changing this situation might have been to resort to private enterprise and to establish institutions which could compete with the official ones. This kind of initiative was effective in Britain in coaxing the old universities into reforms during the nineteenth century. But the governmental monopoly of higher education and science in France was too strong and too comprehensive to allow private initiative the necessary scope which would have enabled it to compete effectively.[39] Such institutions as the *Ecole centrale des arts et manufactures* (established in 1829) or the various private study groups and schools in other fields including the social sciences or even the famous and highly successful Pasteur Institute, remained specialized and isolated efforts, complementing the existing establishments rather than exerting pressure on them.[40]

This situation, where initiative in organizational change was frustrated or thwarted by a centralized system which made each particular organization a negligible quantity, was responsible for the often criticized individualism, fragmentation, and conservatism of French scientific efforts. Since it was virtually hopeless to try to change anything in the system as a whole or in the structure

---

[37] Liard, *op. cit.*, pp. 294–295, and H. E. Guerlac, "Science and French National Strength," in E. M. Earle (ed.), *Modern France* (Princeton: Princeton University Press, 1951), pp. 86–88.

[38] Theodore Zeldin, "Higher Education in France, 1848–1940," *Journal of Contemporary History*, (July 1967), II:77–78.

[39] The centralization of the system has prevented even the most able people from thinking in terms of individual institutions. Even such an outstanding politician of science as Victor Duruy was convinced that the system as a whole was sound and needed only more support for research. See Liard, *op. cit.*, pp. 287–288.

[40] Prost, *op. cit.*, pp. 302–305, and H. E. Guerlac, *op. cit.*, p. 88.

of individual institutions through concerted action by those immediately concerned, the best strategy for the individual scientist was to pursue his own ends "egotistically." He worked as an individual and tried to further his own purposes. The individualistic isolation of scientists from each other provided a parallel to similar phenomena generated by the French political and bureaucratic systems in civic affairs and in many work situations.[41]

This state of affairs imprinted on French science a distinctive characteristic. During the second half of the nineteenth century, scientific work started growing in scale and came to be based increasingly on cooperation and division of labor. Scientists in different fields and in different institutions came to regard themselves increasingly as members of professional communities pursuing common purposes and defending common interests. In France this development was greatly inhibited by the structure described above. This probably had a directly detrimental effect on the quality of scientific work. In addition, it contributed to the relative isolation of French scientists from the international scientific community of which they had been the center early in the nineteenth century (i.e., at the time when scientists everywhere worked as isolated individuals). Elsewhere scientists started forming schools and working in groups. In France, with few exceptions, they went on working as individuals, training their successors as personal apprentices or not training any successors at all.

The Conditions of Reform in France

Under these conditions, changes in French scientific organization occurred in a different way from that in the other scientifically important countries. In the latter, changes were instigated either by the competitive initiative of various independent universities and other institutions, or by the pressures and policies of scientific elites acting as the representatives either of the scientific community as a whole (e.g., the Royal Society in England), or of formal and informal associations of scientists and scientific institutions, as in the United States. In France, innovations occurred not as the result of horizontal combinations of scientists or scientific institutions, but as the result of vertical combinations of individual scientific entrepreneurs or scientific cliques —usually identified with political tendencies—on the one hand, and individual administrators and politicians on the other. To such a short-lived constellation was due the foundation of the *Ecole pratique des hautes études* by Victor Duruy during the last years of the Second Empire. Only rarely did these situations last long enough to leave time for the fulfilment of programs of comprehensive reform. Such a relatively long period did occur between 1879 and 1902. A representative group of scholars and scientists led by the historian Ernest

---

[41] Michel Crozier, *The Bureaucratic Phenomenon* (Chicago: University of Chicago Press, 1963), pp. 214–220; for a detailed description of the problem in relation to science, see Zeldin, *op. cit.*, pp. 67–68, and R. Gilpin, *France in the Age of the Scientific State* (Princeton: Princeton University Press, 1968), pp. 107–108.

105

Lavisse and the chemist Berthelot (the latter also served as Minister of Education from 1886 to 1887), supported by Alfred Dumont and Louis Liard (directors of higher education from 1879 to 1884 and 1884 to 1902, respectively), tried to reform the French faculties on the model of the German universities. Although they did not succeed in this, they expanded the whole system very considerably—the number of professors in France increased from 503 in 1880 to 1,048 in 1909 and then remained nearly static until the 1930s—and they raised its standards.[42] The university structure established by these reforms remained virtually unchanged until 1968.

During the period from 1879 to 1902, and then again in the 1930s under the *Front Populaire* government and after World War II (the *Centre National de la Recherche Scientifique* was established in 1936 and subsequently expanded) French science policy was conducted somewhat similarly to British science policy. It was inspired by a fairly representative informal elite of scientists and intellectuals and carried out by sympathetic governments. But France, unlike Britain, possessed no organizational infrastructure for this movement. The influential scientists were individuals who had mutually congenial outlooks in a situation in which general political trends were favorable to science. These were periods in which there was a greater than usual consensus of a liberal-socialist tone, infused with scientism and favorable to science. But there were no central bodies such as the Royal Society and the Athenaeum where a common point of view could be formed and promulgated and no universities which commanded deep institutional loyalty. Nor did France possess an intermediary body such as the University Grants Committee which could consolidate these elite groupings beyond the duration of politically favorable periods. The French scientific elites were always politically tinged; cooperation within them always involved some tension and was unstable. As soon as the politically favorable constellation passed, as a result of a change of government, or perhaps even of a mere change of ministers or directors of higher education, the elite group was in danger of dissolving into political factions which used the various scientific institutions as a base for individual or clique activities rather than acting in the interests of the scientific community as a whole.[43]

Under these conditions, such continuity of action as existed has been ensured neither by the continuity of an elite, nor by the continuity of independent scientific organizations. The stability of the system—like that of many other things in France—has rested primarily on the central bureaucracy. Between this bureaucracy and the individual scientist there have been no significant intermediary organizations, only shifting cliques. As a result, the system has been ill suited for ventures involving flexibility and cooperation. It has only

[42] Guerlac, *op. cit.*, pp. 83, 88–105, and Prost, *op. cit.*, pp. 223–224, 234.
[43] This apparently happened as a result of the Dreyfus affair, which polarized political passion; see Clark, *op. cit.*: the general instability of conditions comes through clearly in Zeldin, *op. cit.*, pp. 53–80, 69–80, and in Gilpin, *op. cit.*, pp. 112–123.

106

been possible to devise within it a variety of strategies ensuring careers, and some, but rarely sufficient, means for research.

These are the factors that explain the relative inefficiency of the French system, as compared to the British, in keeping pace with scientific centers in Germany and the United States. The inability to compete successfully is not a result of any lack of motivation to excel in science. This motivation has been institutionalized in French society, and it has brought forth brilliant scientists as well as imaginative policies to further science. But because of their dependence on passing political constellations, the policies designed to improve French scientific research and training have had little continuity. Moreover, the absence of independent scientific organizations commanding the loyalty of scientists and encouraging their cooperation has inhibited the growth of up-to-date patterns of scientific work. Both the dependence of the system on the vicissitudes of politics and its organizational rigidity derive from its centralized bureaucratic organization.

german scientific
hegemony and the
emergence of
organized science

seven

The Transformation of Scientific Work
in the Nineteenth Century

The transformation of science into a status approaching that of a professional career and into a bureaucratic, organized activity took place in Germany between 1825 and 1900. By the middle of the nineteenth century, practically all scientists in Germany were either university teachers or students, and they worked more and more in groups consisting of a master and several disciples. Research became a necessary qualification for a university career and was considered as part of the function of the professor (although not an officially defined part). The transmission of research skills usually took place in university laboratories and seminars, rather than in private. Finally, during the last decades of the nineteenth century, research in the experimental sciences became organized in so-called institutes, which were permanent bureaucratic organizations usually attached to universities that possessed their own plants and scientific and supporting staffs.

Systematic training and division of labor now became important factors of scientific growth in addition to, and independent of, spontaneous interest and popular support of science. Thus the place of science and its relative strength in a given country when compared with other countries could not be fully explained any longer by the strength of the scientistic movement or the extent

108

of the institutionalization of science. The effectiveness of the universities and other research organizations, and the status of science vis-à-vis other disciplines within the universities (which was not necessarily the same as outside the universities) became independent determinants of scientific growth. In other words, a country where the scientistic movement was weak could still become a scientifically leading country as a result of support given to a relatively independent and socially insulated system of higher education and research. That support was given on grounds unrelated to the acceptance of science as a value in itself. The purpose of this chapter is to explain how this transformation came about.

The Social Situation of German Intellectuals

The first step toward this transformation was the establishment in 1809 of a new type of university—the University of Berlin. The new university was imitated within a short time by the whole system of German language universities.[1] These innovations, like those which had occurred in France about 1800, were initiated by intellectuals, and their original shape was determined by the needs and the ideas of this group. The initial differences between the new French and German systems were due to the differences between the composition and character of the intellectual classes that in turn were rooted in the altogether different class structures of the two countries.

By English or French standards, Prussia had been a backward country even in the eighteenth and early nineteenth centuries. Its middle class was small and lacked political power, and many of the social classes, including a majority of the bourgeoisie, were traditional.[2] The rulers of the kingdom, however, had been very successful in the creation of a well-organized army and civil service, which were completely responsible to the king. The rulers began fostering commerce, industry, and education at all levels with considerable success, without abandoning any of their traditional prerogatives. As a result, groups of young people emerged who were educated according to the ideas and the ideals of the French Enlightenment in a country which had still been feudal on the local level and where there was no plurality of politically important socioeconomic groups. In fact there were no groups at all that were sufficiently wealthy and important to gain any independence from the ruler and his bureaucrats. Only in the religious sphere was the situation comparable to that prevailing in the West. It was comparable in the sense that there existed genuine plurality and, therefore, a preparedness to accept a neutral sphere of intercourse about philosophical

[1] Franz Schnabel, *Deutsche Geschichte im neunzehnten Jahrhundert*, Vol 2 (Freiburg, Breisgau: Herder Bücherei, 1964), pp. 205–220.

[2] Henri Brunschwig, *La crise de l'état prussien a la fin du XVIIIe siècle et la genèse de la mentalité romantique* (Paris: Presses Universitaires, 1947), pp. 161–186; Werner Sombart, *Die deutsche Volkswirtschaft im neunzehnten Jahrhundert*, 5th ed. (Berlin: Georg Bondi, 1921), pp. 443–448.

and artistic matters where people of different religious persuasions could pursue common interests.[3]

As a result, the practical economic and political concerns of the English-French philosophers had limited meaning for the newly "Westernized" German intellectuals. Because there were no important social groups pressing for political freedom and social equality, there was not much interest in the differences between the various models of society developed by the English and French political thinkers. Similarly, in the absence of a powerful entrepreneurial class, there was no interest in the applications of scientific models to political economy. Such models were relevant in an economy where decisions were made by a large number of individuals working independently of each other, but not to economies where the framework of the economy was fixed by despotic rulers and tradition.[4] The English and French ideas were applicable to German society only in their secular attitudes toward religious differences. The idea of a secular culture was a potentially feasible program in Germany (as in the Western countries), and it enjoyed the support of important social groups.[5]

Hence the relationship between science, scholarship, and philosophy became quite different in Germany. The development that started with Rousseau's questioning of the relevance of science to moral problems, and which led him in search of alternative bases for the development of a secular philosophy, was more relevant to the German situation than empirically oriented political science and economy. Thus romanticism and idealism, which remained as intermittent and secondary philosophical trends in France and Britain, became the major trends in Germany. And for the same reason, German rather than English-French philosophy served as the principal model for the philosophers of Eastern European nations during the nineteenth century.[6]

Furthermore, there were differences between France and Germany even within the framework of the new ascientific philosophy. The alternatives to the scientific model in Germany were not the concepts of general will and unspoiled human nature. These concepts were still related to a situation where social reform was a central issue of public debate. They were used as alternatives for the science-based philosophies, but their objective was social and educational

[3] Jacob Katz, *Die Entstehung der Judenassimilation in Deutschland und deren Ideologie* (Frankfurt: a/M, 1935).

[4] On the backwardness of German economics see H. Dietzel, "Volkswirtschaftslehre und Finanzwissenschaft," in W. Lexis (ed.), *Die deutschen Universitaeten*, für die Universitätsausstellung in Chicago, Vol. I (Berlin: A. Ascher, 1893). Steps were taken only in 1923 toward the updating of economic studies at the universities. See Erich Wende, *C. H. Becker, Mensch und Politiker* (Stuttgart: Deutsche Verlags-Amstalt), 1959, p. 129.

[5] Katz, *op. cit.*; Schnabel, *op. cit.*, p. 206

[6] The most important effect of romanticism in the Western countries was on literature. See Bertrand Russell, *A History of Western Philosophy* (New York: Simon and Schuster, 1945), pp. 675–752. On the overwhelming German influence on Russian philosophy see Alexander von Schelting, *Russland und Europa im russischen Geschichtsdenken* (Bern: A. Francke, 1948), and Schnabel, *op. cit.*, Vol. 5, pp. 186–194.

german scientific hegemony and the emergence of organized science

change. The rejection of the empiricistic framework in the French school was only partial. Some French philosophers rejected the validity of the Newtonian model for human affairs, or even for science in general, but they were still concerned with empirical problems of political change and scientific inquiry.

In Germany, philosophy took a much more abstract turn. The main concerns of German philosophy were aesthetic self-expression of the individual and of the nation through its unique culture, the establishment of a systematic theory of metaphysical knowledge, and moral values based on intuition and speculation. This change of emphasis and interest reflected the fact that in Germany the intellectuals could have no pretensions for political leadership. They had to concentrate, therefore, on spiritual matters. In these they could be assured of an audience, since, as has been pointed out, religious pluralism provided a sympathetic social background for the philosopher's search for a secular, spiritual, and moral culture.

These purely spiritual concerns were reinforced by the social status of the German intellectuals. Unlike the intellectuals in France, they were usually not well-to-do people with independent incomes, nor were they generously supported by rich patrons. They came from the modest middle-class background. A generation before, such intellectuals would probably have become clergymen engaged in preaching and teaching. By the beginning of the nineteenth century, however, they did not want to become clergymen any longer, and teaching itself was a very poor career. Although some German universities in the eighteenth century had employed up-to-date scientists and scholars among their teachers, there was still a great deal of clerical supervision of the universities. The main problem, however, was the status of the university teacher in the arts and sciences. These teachers usually taught only students studying for the lower academic degrees, and the philosophical faculty was subordinate in status to the faculties of law, theology, and medicine. Correspondingly, the status and the income of teachers in the philosophical faculty was much lower than in the higher faculties.[7]

This frustration of the German intellectuals by the universities was made worse by the preference for French scientists and philosophers among the ruling class. In the eighteenth century German academies invited foreign scholars in preference to Germans. Maupertuis was the president of the Berlin academy, and his adversary, Voltaire, was the most respected philosopher at the Prussian court under Frederick the Great. Thus not only were the opportunities for obtaining support and recognition outside the universities (in academies and through private patronage) few in Germany compared with Britain and France, but there was also a discrimination against German intellectuals. By the beginning of the nineteenth century, however, the situation of the German intellectual had improved although he still felt that he had to assert himself against the French. Thus the dissatisfaction with the backward state of the universities at the end of the eighteenth century, which was common to both Germany and

[7] Brunschwig, *loc cit.*; Schnabel, *op. cit.*, Vol. 2, p. 182.

german scientific hegemony and the emergence of organized science

France, led to different reactions in both countries. In France the intellectuals, who were led by the scientists, agreed to the abolishment of the universities and eventually accepted the *Grands Ecoles* and the specialized faculties as a replacement for the universities. In Germany the intellectuals, led by philosophers and humanists, resisted the attempt to follow the French example in the reform of higher education which was advocated by the "enlightened" civil servants. They agreed with the latter that the university had to be basically reformed. But the replacement of the universities with specialized high schools would have threatened the existence as well as the cultural mission of the German intellectuals. What they were interested in was raising the status of the universities and of the philosophical faculties within the university to the level of the academies.[8]

This problem was common to some of the scientists and the humanists, both of whom either worked at universities or aspired to work there. A further ground for a feeling of common interest between German humanists and scientists was the tendency among humanists to be much more scientific than elsewhere. This tendency might have been a result of the above-mentioned fact that in Germany, as in other culturally peripheral countries such as Holland, the Scandinavian countries, and Scotland, the scientists did not secede from the universities in the seventeenth century as they did in the important centers such as France and England. In any case, humanists in Germany increasingly modeled their behavior after that of the scientists. They considered cultural phenomena, like history, literature, and language, as empirically existing subjects and regarded philological inquiry as a method of empirical scientific research. Humanities ceased to be cultivated as instruments of aesthetic and moral education that were studied in order to shape one's character, style, and thought. Instead, they were considered subjects to be understood, just like natural phenomena. They were to be approached objectively and in a value-free manner. In fact, the humanities were considered as empirical sciences, and, at times, even as models for empirical research.[9]

Of course this approach to culture was not unknown elsewhere. But the distinction between the "scientific" and the educational approach to the humanities had nowhere been drawn so sharply as in Germany (and, subsequently, the rest of Central and Eastern Europe), and the identification of the study of the humanities with that of the sciences as fields of empirical scientific inquiry had nowhere become so programmatic.

This identification became the basis of the common claim of the scientists and the humanists for demanding equal status for the philosophical faculty with the professional faculties. Eventually the claim led to the transformation of the university into a scientific institution whose members were engaged in creative research. The successes of natural science and the superiority of exact

[8] Rene König, *Vom Wesen der deutschen Universität* (Berlin: Die Runde, 1935), pp. 20 ff., 49–53; Schnabel, *op. cit.*, Vol. 2, pp. 198–207.
[9] Schnabel, *op. cit.*, Vol. 5, pp. 46–52.

philology over the older methods of studying the Bible and the classics were the main justification of this intra-academic fight for the reform of the university. In a way, this was a repetition of the seventeenth-eighteenth century fight of the new experimental and empirical method of inquiry against the old scholastic approach and the time-honored but unscientific professional traditions. There was, however, a sociologically decisive difference: because of their common interests in the advancement in the academic scale of prestige, humanists and scientists made the common, scientific features of their methodology the basis of the definition of the two fields. This severed the links of both fields with their applications. Natural scientists separated themselves from technology and practical social philosophy; humanists ceased to take interest in creative writing and moral education. The severance of science from its uses in technology, which was avoided in France, became part of the intellectual program in Germany. This separation was accompanied by a parallel severance of the link between the humanities and their educational applications.

Distinguishing the scientist's role from that of the practical social philosopher had been a discernible tendency among some scientists in the seventeenth and eighteenth centuries in order to emphasize the neutrality of science. But the inclusion of the humanists and the exclusion of the technologists in the definition of the role only made sense under the conditions prevailing in the German and other similarly structured universities. Had scientists been important gentlemen as in England and France, they would probably not have been enthusiastic about being identified with humanist scholars who were mainly college teachers engaged in inquiries of prescientific or ascientific cultural contents. Even in Germany at least part of the natural scientists were more inclined to accept the English-French definition of their roles. Some of them favored the French type of specialized schools; in any event, there were no natural scientists among the great figures advocating the new type of university. But the natural scientists were only a small group among the intellectuals, the bulk of whom were schoolmasters teaching languages and humanities. Their goal was to be recognized as scientists, and therefore, as such, to be engaged in a value-free intellectual pursuit not controlled by state and church authorities. They also wanted recognition of abstract, nonutilitarian science (it had to be nonutilitarian, since philology and history had no practical use) as a major concern of higher education equal in status to teaching for the learned professions. All this was in sharp contrast to the French reform movement, which was spearheaded by scientists and scientistic philosophers.[10]

[10] Out of thirty-one people mentioned by Liard as having taken part in the initiation and planning of the reform of French higher education during the Revolution, at least twelve were well-known scientists, two were philosopher-scientists, one was an economist, one (Sieyès) can be described mainly as a political ideologist, and only one was a well-known literary man (Auger). The rest were politicians and educators. The most influential thinker on the subject was Condorcet, and the most distinct group consisted of the scientists (Lakanal, Fourcroy, Carnot, Prieur, Guyton de Morveau, Monge, Lamblardie, Berthollet, Hassenfratz, Chaptal, and Vauquelin) See Louis Liard, L'Enseignement Supérieur (Paris: Armand Colin, 1894),

The relationship between the philosophical movement and this scientific-scholarly community in Germany was as ambivalent as that between scientistic philosophy and the scientific-technological community was in England and France. That is, there was much doubt on the part of the experts about the philosophers—who were engaged neither in exact nor in empirical research. At the same time there was much common interest and overlap between the two groups.[11] Furthermore, German philosophers asked as legitimate questions about the logic of the scientific study of the humanities and the cultural implications of the systematic study of art, literature, and history, as the British-French philosophers had asked about the logical foundations of natural science and its implications for society.

In this new conception of philosophy, natural science no longer served as the model of intellectual inquiry. Philosophy was concerned with the creation of a substitute for a comprehensive world view of the kind offered by religion. Natural science had its place in this philosophy as an important, but not the most important segment of human knowledge. In fact, it was considered as a poor third after speculative philosophy and the humanities which, of course, dealt with spiritually more important subjects. Neither natural science, nor any other kind of philosophically important knowledge, had to be directly useful for political or economic purposes either in the short or in the long run. Learning and knowledge were ends in themselves. Their importance derived from providing a spiritual justification for society and from their educational effects of shaping the mind.

This view of philosophy and its relation to the humanities and the sciences—all of which were considered as science—was actually a return to the Greek philosophical conception. In this view the role of the specialized natural scientist became as ambiguous and potentially as marginal as it had been in antiquity. This change of perspective was reflected in the structure of the faculties. The sciences and the humanities were both to become parts of the philosophical faculty, while in the French revolutionary schemes philosophy and science were separated from the humanistic subjects.[12]

Thus there occurred a complete reversal in the main trend of philosophical thought. The methods of acquiring logically correct and empirically valid

---

Vol. I, pp. 117–311. In Germany the most important group of intellectuals participating in the process were the philosophers (Fichte, Schelling, Schleiermacher) and the philologists (Wolf and Humboldt). Their opponents, who preferred specialized institutions, were educationists and civil servants (see König, *op. cit.*, and Schnabel, *op. cit.*, Vol. 2, pp. 173–221). The few scientists who took part in the discussions also favored the specialized institutions (schools such as medical, mining, etc.) where most of them taught. See Helmut Schelsky, *Einsamkeit und Freiheit; Idee und Gestalt der deutschen Universität und ihrer Reformen* (Reinbeck bei Hamburg: Rowohlt, 1963), pp. 36–37.

[11] Schnabel, *op. cit.*, Vol. 5, pp. 49–52.

[12] See Liard, *op. cit.*, Vol. I, pp. 396–463, about the different plans and blueprints for the division of the faculties under the Revolution.

knowledge of nature and society ceased to be the main theme of philosophy. Instead, its paramount scientific concern now became the study of culture, namely the immense variety of human self-expression.

This reversal of interest took place within the same universe of discourse as British-French philosophy. Its basic questions were logically related to the questions of that philosophy. Sooner or later the nature of the knowledge of cultural phenomena had to be raised in the framework of the new secular philosophy. But in Britain and France the question was usually avoided, because the subjectivity of cultural values seemed to preclude an objective philosophical approach. The raising of this question in Germany reflected the relative lack of interest in natural sciences and in empirical social thought, and the interest in a secular substitute for a quietistic religiousness and in spiritual culture. This choice of philosophical interest implied a breach with the scientific tradition which avoids questions that do not have universally valid solutions. Diderot and perhaps even Rousseau and Kant, who also asked such questions, still shared the belief in the possibility of finding answers that could be universally validated by experiment or experience. But Fichte, Schelling, Hegel and their contemporaries believed that they had found the key to complete and final knowledge in their own intuition and did not feel that this knowledge needed any further validation. All that had to be done was to interpret everything known in the light of this new knowledge. Their advocacy of the creation of a new type of university, autonomous in setting its own aims and engaged in the pursuit of pure knowledge, was based on these views. According to this concept, philosophy (which embraced all knowledge) was more important than any other study. Everything was subject to the critique of philosophy, while philosophy could not be tested by anything else.

Thus it was even more pronounced in the German case than in the French that the establishment of the new system of higher education was a response to the needs and pressures of intellectuals in general rather than to those of the expert scientists. Unlike the situation in France, however, in Germany the most influential members of this intellectual group were ascientific philosophers and, next to them, the scientific humanists (the latter were probably the majority). As a result, the entire reform was rooted in a conception of science that included speculative and nonmathematical philosophy, as well as the humanities studied according to the philological method. As has been pointed out, this conception implied a redefinition of the social functions of science and scholarship and an even more drastic revision of the social function of philosophy.

The extreme views of the idealist and romantic philosophers were not shared by Humboldt, who was the most influential single person in the establishment of the University of Berlin. These views were also opposed by some of the humanistic scholars who regarded themselves as empirical scientists. Nevertheless, these philosophies had a decisive influence in the universities. The opposition of the empiricistic humanists to the philosophers was restricted

115

to the professional interests of the former. They did not want to be taught how to consider history or jurisprudence by the philosophers who claimed competence in everything. But at least tacitly they accepted the philosophical view of the superiority of the study of spiritual culture to natural science and the nonutilitarian and nonempirical view of higher education according to which every education, including that for the practising professions, had to include basic training in some—preferably humanistic—cultural content. Theologians were supposed to study Hebrew and Greek philology; lawyers historical and philosophical jurisprudence; and doctors *Naturphilosophie*. Thus, originally the spirit prevailing in the new universities was more a revival of the spirit that had prevailed in the Greek schools of philosophy than an attempt to base education on modern science.[13]

Reform of the German Universities

The triumph of these ascientific (and at times antiscientific) views can be explained only by the special circumstances prevailing in Prussia during the crucial years preceding the establishment of the University of Berlin. Originally the government circles favoring the reform of higher education were influenced by French ideas and favored the Napoleonic model over the establishment of a university with a philosophical faculty as its center. The turning point in favor of the ideas of the philosophers came under the Napoleonic Wars. The advocacy of the German brand of philosophy, unheeded before by the upper classes, now became acceptable to all. There was a feeling that the real strength of the nation was in the realm of spirit. Indeed, after their defeat by Napoleon, the Germans could find comfort only in the unprecedented flowering of national philosophy and literature. For the first time, German philosophers became important public figures in their own country and their advice was heeded, especially in matters of education.[14]

Thus, unlike in France, the support given to the new universities in Prussia was not the result of the acceptance of the scientistic philosophy by rulers. And the purpose of the reforms was not the creation of a society where the scientific approach would prevail in government and economy. Support was forthcoming because of the acceptance of a new speculative philosophy that extolled an ascientific idea of a nationalistic philosophical, literary, and historical culture that was believed to be superior to everything else in the world. The

---

13 See Schnabel, *op. cit.*, Vol. 2, pp. 219–220, about Humboldt's refusal to co-opt any scientific specialists in the scientific advisory committee of his ministry. The members were philosophers, mathematicians, philologists, and historians. He thought that the subjects represented "include all the formal sciences . . . without which no specialized scholarship (*"auf das einzelne gerichtete Gelehrsamkeit"*) can become the true intellectual education. . . ." About Humboldt's (and others') opposition to training people at the university for any practical purpose, see *ibid.*, pp. 176–179, 205–219. For the similarity of the new conception of higher education to ancient Greek ideas see *ibid.*, pp. 206, 217.

14 *Ibid.*, p. 204.

116

university representing this philosophy was granted autonomy. This support, however, was not tantamount to the acceptance of free inquiry as an independent and socially valuable function. Rather, there was a presumption of preestablished harmony between the new philosophy and the interests of the state, somewhat in the same manner as there had been an assumption of such harmony between the church and the state.

Although there was some opposition to this view among the humanist scholars at the universities, the natural sciences had not been favored in the new universities. Many chairs in the natural sciences were filled with adherents of the romantic *Naturphilosophie* which opposed mathematics as well as experiment.[15] And the conception of professor as a person charged with the presentation of an original and presumably complete and closed view of a whole scientific discipline was fitted to the occupation of philosophical system builders, or students of closed cultural contents, rather than to that of the empirical scientist working as a member of a constantly changing research front.

This problem was understood by Humboldt and some others who participated in the formation of the new type of university. Therefore, in the role of the *Privatdozent,* they tried to establish structural safeguards to ensure the place and maintain the autonomy of researchers in the universities. This safeguard was not sufficient, however, and during the first twenty years of its existence the new German university did perhaps more harm than good to natural science (social science had been suppressed even longer). There was a tendency in the German universities from about 1810 to 1820 to deny the distinctness of the role of the empirical-mathematical scientist, which had been created in England and France during the seventeenth and eighteenth centuries. And as far as the diffusion of the scientific approach to social and moral philosophy was concerned, the trend was completely reversed in Germany. Nevertheless, by the 1830s the tide had turned and there ensued a flowering of the natural sciences and of the experimental approach in general at the universities.[16] By the second half of the century a serious trend toward the diffusion of this approach to psychological and social phenomena began. Our discussion will now show how this occurred.

Organizational Structure of the University

It is interesting to note that those who wrote about the German universities and scientific life in general did not usually take into consideration the fact that the immediate result of the establishment of the new German university was the decline of empirical natural science. Hence the attribution of the eventual scientific productivity of the German universities to the philosophical ideas that predominated at the time of the reform is not

[15] *Ibid.*, Vol. 5, pp. 207–212, 222–238.
[16] *Ibid.*, pp. 238–276.

supported by any evidence.[17] Philosophical idealism and romanticism might have inspired the imagination of a few scientists. But inquiry for its own sake in these philosophies was conceived mainly as speculation. The original structure of the professorial role was ill-suited to empirical science. The rise of empirical science starting from the late 1820s (due to the work of such pioneers as Liebig, Johannes Müller, and their disciples) was not a result of the new university, but of a conscious revolt against its philosophy and of an important, although not deliberate and conscious, modification of its structure.[18]

The superiority of the German to the French system is to be sought, therefore, in the capability of the German system to change itself according to the needs and potentialities of scientific inquiry in spite of the wrong ideas (from the point of view of empirical science) of the university's founders. By contrast, institutions in the French system, even if originally well conceived, were incapable of adapting themselves to changes.

This capability of the German universities to change themselves could be situated in the internal organization of the university, in the system of the German universities as a whole, or in the interaction of the two. The existing literature usually stresses the importance of the internal organization, which reflected the philosophical ideas. However, this discussion will try to show that the decisive condition was the way the system worked as a whole.

The arguments about the superiority of the internal organization of the university stress two features: (1) academic freedom and self-government, and (2) the definition of the two principal academic roles, those of the *Privatdozent* and the professor. The former, it is argued, ensured that decisions about academic matters were made by experts who were motivated first and foremost by scientific interests and who had the knowledge to act effectively; while the latter, namely those who met the requirements for the qualification of *Privatdozenten* (out of whose ranks the professors were usually chosen) made research an integral part of the academic role. Our discussion will now show to what extent this is a satisfactory explanation of the greater adaptability of the German academic institutions.

Let us start with the problem of academic freedom. The changeover to an increasingly professionalized pattern in science in France and Germany raised the problem of how to ensure that regular employment in governmental bureaucratic frameworks would not interfere with the freedom and spontaneity of scientific creation. In addition, in Germany and most of the other countries of Central and Eastern Europe, there was the problem of safeguarding freedom of inquiry. This freedom, of course, was not an academic problem in nineteenth

---

[17] The originator of the view about the perfection of ideas and arrangements at German universities was probably Victor Cousin. They were most widely propagated by Abraham Flexner, *Universities: American, English, German* (New York: Oxford University Press, 1930). For the general view of German philosophy as the root of German "progress" see Elie Halévy, *History of the English People (Epilogue: 1895–1905, Book 2)* (Harmondsworth: Pelican Books, 1939), pp. 10–13.

[18] See footnote 16.

german scientific hegemony and the emergence of organized science

century France, because freedom of speech was considered as part of the rights of every citizen.[19]

In Germany and elsewhere, however, there was neither freedom of speech, nor social equality. Nor was there any powerful support for these freedoms. Science had to be accommodated in an inimical environment, and special safeguards had to be devised for securing its freedom. This was done in two steps. First, an organization had to be devised that could be given special privileges of freedom without creating a precedent for the granting of democratic freedom to the people in general. Second, this organization had to be set up in such a way as to prevent its becoming the kind of autocratic and hierarchic bureaucracy which had been the only one known in Europe at that time and which could not conceivably be suited for the requirements of creative research. The prevailing paradigms of bureaucracy were the military, the civil service, and the Catholic and Lutheran churches, and none of these seemed suitable for scientists. Experience such as existed in the Scottish Presbyterian church, the important noncomformist sects, and the innumerable societies and associations in Britain sharing a variety of political, administrative, and judicial responsibilities, were hardly known and of little importance on the Continent.

The difficulty of fitting scientists into governmental bureaucracy was evident to all the reformers of higher education. The solutions of the problem were based on three premises: First, it was assumed that the scientist worked as an isolated individual and not as a member of a unit. Second, his contractual obligations were strictly limited to teaching and examination for certain degrees conferring full or partial qualifications for the liberal professions, secondary school teaching, or the civil service. The curricula, contents of the courses, and hours of teaching were minimally defined, and it was assumed that the teacher's free time and/or the undefined part of his teaching were to be used for research and original writing and lecturing. The precondition of this latter understanding was that academic teachers should be outstanding scientists. This was to be ensured by more or less successful procedures of selection and appointment. Third, scientific research was not to become a career for which one was regularly trained, but a calling for which one prepared and devoted oneself privately. Those who were exceptionally able and lucky were then rewarded publicly by positions ensuring income. But these positions were regarded more as honors than the culmination of reasonably calculable careers.

Scientists, therefore, who derived their income either from being inspectors

---

[19] This is not to say that there were no conflicts about the political and religious pronouncement of the academics in France. But these were the result of the open involvement of the *université* (which also included secondary education) in politics. Even under these conditions of the politization of the whole educational system, repressive measures (such as dismissal) were applied only in secondary schools and very rarely—and then only for short periods of time—in higher educational establishments. See Paul Gerbod, *La condition universitaire en France au XIXe siècle* (Paris: Presses Universitaires, 1965), pp. 103–106, 461–474, 482–507, 555–563, and Albert Léon Guérard, *French Civilization in the Nineteenth Century: A Historical Introduction* (London: T. Fisher Unwin, 1914), pp. 230–237.

of education in France or university professors in Germany, were not actually paid for research, but for some work which could be defined in a routine bureaucratic manner. Having obtained the reasonably lucrative and not too arduous appointment, they were supposed to engage in science as free gentlemen of leisure according to the old amateur pattern.

The problem, of course, was much easier in France because scientists were employed there in a variety of capacities and institutions. As some of these positions had nothing to do with teaching (which was not taken too seriously anyway) and few of them were exclusively given for research (some were only sinecures), there was no real danger of the systematic bureaucratization of scientific inquiry. If bureaucracy became too bad in one place, there were always several others where scientists could find freedom. Or, in the worst case, they could advocate the setting up of an entirely new kind of institution.[20] In Germany, however, where the institutional privileges were the only basis of scientific freedom, and where scientists had little influence on government before the end of the nineteenth century (and then still not too much), careful thought had to be given to this problem.

In the absence of a wealthy and free middle class and/or strong liberal parties supporting the cause of science against governmental despotism, the only available social device for safeguarding the freedom of science was the old academic corporation. The attitude toward this, however, was ambivalent. On the one hand "enlightened" public opinion (the intellectuals and government circles), regarded the old university corporations as reactionary bodies that were responsible for the decay of the university. On the other hand, the new romantic anti-French spirit emphasized the virtues and the original German character of corporate self-government.

The solution to the dilemma was to transfer to the state the functions of financial supervision of the universities, of responsibility for part of the examinations qualifying for professional practice, and of the appointments to the chairs. This latter function, however, reverted in practice to the academic corporation, although the state preserved its ultimate rights. The university senate remained in charge of all the academic affairs. Thus the reformed and enlightened state was to assume responsibility for the furtherance of science and to prevent the development of guildlike rigidities, while the academic corporation was to counteract any despotic tendencies of the state and safeguard the freedom of the individual researcher.[21] The corporate structure was not chosen for its flexibility and efficiency, and, indeed, as will be shown below, it was neither flexible nor efficient. It is very doubtful, therefore, that academic self-government contributed positively to the adaptability of the German system. However, it probably con-

[20] Maurice Crosland, *The Society of Arcueil: A View of French Science at the Time of Napoleon* (London: William Heinemann, Ltd., 1967), pp. 228–229, shows that teaching was considered only as a means, and only one of the means, for the material support of scientists.

[21] Schnabel, *op. cit.*, Vol. 2, pp. 211–215.

tributed to this end negatively by jealously protecting the right of its members to do as they pleased, and to innovate and engage in new kinds of ventures in their own fields as long as this did not interfere with the interests of the others. This activity, however, depended entirely on the motivations and qualities of the membership. Corporate freedom lent itself equally well to the protection of abuses committed by individuals or the safeguarding of common vested interests. Thus the effectiveness of the system depended on (a) the quality of those recruited and (b) either the absence of collective vested interests inimical to science, or the existence of some countervailing force capable of neutralizing corporate defense of selfish interests by the universities.

Indeed great attention was paid to ensure the high quality of appointments. The requirement for academic appointment was the *Habilitation*, which was supposed to be an original contribution based on independent research. This requirement has been favorably contrasted to the corresponding qualification in France, the *aggrégation*, a difficult competitive examination held every year in each academic field. The *Habilitation* was thus supposed to ensure that the appointments to professorships would be competent and highly motivated researchers. There were also some arrangements to create an independent countervailing force to the corporation of university professors. One of these was the institution of the *Privatdozentur* (private lecturership). According to this arrangement, those who obtained a *Habilitation* had the right to lecture at the university even if not elected to a chair (they did not have salaries, however, and obtained only attendance fees paid by the students who chose their courses). From this community of scholars and scientists in different fields those who excelled were to be elected as professors. Election conferred on the professor special emoluments and honors. Otherwise, a chair was not supposed to change the conditions of the professor's work, nor was it supposed to confer on him any authority over the *Privatdozenten*. All remained free and equal as scientists responsible only to their own scientific consciences, and the public opinion of the scientific community and the students.

The other countervailing force was to be the freedom of the students to choose their lectures, to attend or not to attend them, and to transfer credits from one university to another. This was conceived as a system of checks and balances. Any shortcomings of the privileged professors were supposed to be revealed by the independent *Privatdozenten* and the students, who could show their disapproval effectively by moving elsewhere.

In effect, however, these checks and balances were much less than optimal. The fallacy was that the university community was conceived as being identical with the scientific community, which it was not. The university community consists of specialists in a great many different fields. Therefore, professors and *Privatdozenten* of a university did not represent any effective scientific community (if this latter is defined as a group with a shared competence harnessed to the exploration of fields of common interest). They might have all had the same values or ultimate goals, but values and ultimate goals that are not

121

operational are useless for the formulation of criteria and norms of action. Ultimate values do not help to determine the value of a scientific contribution, or the merits of the contributors, nor can they serve as a practical guide to the organization of teaching and research in any given field.

The university as an aristocracy of merit which was kept in check and spurned into activity by the competition and the criticism of two lower, but free, "estates," namely the *Privatdozenten* and the students, was, therefore, an idealized image. Neither of the estates shared much common scientific competence. On the other hand they shared common class interests. Each professor in his own field had in fact personal authority over the *Privatdozenten*, or the candidates for the title in his own field, and the same applied to the relationship between teachers and students. From the point of view of scientific competence, professors, *Privatdozenten*, and students in a given field formed a community. But this scientific community was divided by gaps of authority and power. The estates of the university, therefore, were based more on shared authority and power than on shared competence. A *Privatdozent* found little in this structure to make possible the effective exposure of the incompetence or narrow-mindedness of a professor through competition. Neither the rest of the professors, nor the other *Privatdozenten* were really competent to judge, and if it came to conflict, it could be expected that the professors would close ranks.[22] The professors alone conferred the title, and their assembly, the senate, held the key to promotions, new appointments, and the creation of new chairs. Indeed from the very beginning there were conflicts between *Privatdozenten* and professors that were brought to the notice of the ministers of education.[23] The ministers also had to interfere in university appointments in order to override decisions of university senates motivated by prejudice and vested interests.[24] Thus all the ingenious arrangements of the university could not provide the checks and balances for which they were intended. The university community was not identical with the scientific community of competent researchers in a field. And there was no formal structure through which universities could be influenced by such scientific communities (whose members were, of course, dispersed over a country, or, rather, over many countries). Rather, the formal structure of the university counteracted the effective development of such communities by the creation of an invidious power and status gap between the elites of the communities and their other members.

What the formal constitution of the single university did not provide,

[22] For a description of the arrangements of the German universities in the nineteenth century, see F. Paulsen, *The German Universities* (New York: Longmans Green, 1906). For the problem of the dependence of the granting of the *Habilitation* in fact by a single authority in each field in every university and an attempt to remedy the situation, see Wende, *op. cit.*, p. 119.

[23] Alexander Busch, *Geschichte des Privatdozenten* (Stuttgart: F. Enke, 1959), pp. 54–57.

[24] See Schnabel, *op. cit.*, Vol 5, pp. 171–175, 317–327, about a series of most important appointments made against the recommendation of faculties.

german scientific hegemony and the emergence of organized science

however, was provided by the system of the universities as a whole. The condition that counteracted the oligarchic tendencies of university senates was the competition among a great number of universities within the large and expanding academic market of the politically decentralized German-speaking areas of Central Europe. Competition among universities checked the development of oppressive academic authority within the individual universities. As long as these circumstances lasted, a situation existed where effective use of resources could be combined with great freedom of the scientific community.

This freedom of the scientists, which was ensured by a competitive and expanding university system, made it possible for individual scientists to undertake and initiate significant innovations. The universities, as a group or individually, did not develop physics, chemistry, or history. They did not have executives whose function it was to foresee and facilitate scientific developments. This was done by the interactive efforts of physicists, chemists, and historians working at different universities or occasionally at other places. They did so as individual entrepreneurs, or in small groups consisting usually of a master and his disciples. But their work was encouraged and facilitated by the existence of the lively demand of universities for successful researchers. As long as there was a seller's market in science, there was always a university that could be persuaded to adopt an innovation. Academic vested interests often opposed innovations, and, as has been mentioned, many an important appointment was forced on unwilling university senates by heads of the university sections of various ministries of education. The ministers used the residual powers vested in the state to override the decisions of self-governing university bodies.[25] Thus competition among the universities and the mobility ensuing from it created an effective network of communications and an up-to-date public opinion in each field that forced the universities to initiate and maintain high standards. It was the interuniversity networks of communication and public opinion in the different fields, and not the formal bodies of the university, that represented the scientific community. The pressures of this informal community (which arose and gained influence as a result of the working of the decentralized system), rather than the corporate structure of the university ensured that academic policies were guided by the needs and potentialities of creative research.

Emergence of the University Research Laboratory

These opportunities for innovation led toward the emergence of regular training and careers in scientific research. Since there was a regular market for successful researchers in Germany, it paid to invest in research. In England during the first half of the nineteenth century a young

---

[25] *Ibid*. For a systematic analysis of how competition influenced the growth of physiology, see A. Zloczower, *Career Opportunities and the Growth of Scientific Discovery in 19th Century Germany* (Jerusalem: The Hebrew University, The Eliezer Kaplan School of Economics and Social Sciences, 1966).

man could engage in research only if he could afford it as a hobby or if he was devoted to science to an extent that he was willing to face real deprivation for its sake. In France the situation was slightly better. There, able young men who had passed the hurdle of several more-or-less difficult and irrelevant examinations could obtain jobs where they could afford to pursue research part-time with some prospect of gradual advancement to positions permitting more and more free time for research.

Both in England and in France, however, the initial opportunity to go into research was the result of means or positions attained for other reasons. Once a person could afford to do research and turned out to be successful, he could use his fame for obtaining further facilities and means of livelihood for the pursuit of his interest. But in Germany, where there was a regular market for researchers, it was possible to make a more-or-less realistic guess about employment opportunities for researchers, go into research straight away at the university, and regard the four or five years spent working on one's thesis and *Habilitationschrift* as an investment toward entrance into a reasonably well-paying and highly interesting position.[26]

It was this gradual transformation of research into a career that enabled the universities to realize the ideal that teachers should also be creative researchers. Those who wanted to go into research were interested in obtaining training for it. This situation made it possible for a teacher to use his academic freedom for the purpose of concentrating much of his teaching efforts on the scientific training of the few would-be researchers. He could then use his bargaining power and that of his students (who were free to move and transfer credits to any other German-language university) to obtain laboratories and other facilities for research.

As a result, the laboratories of some German universities became the centers and sometimes virtually the seats of world-wide scientific communities in their respective fields starting about the middle of the nineteenth century. Liebig at Giessen, and Johannes Müller at Berlin were perhaps the first instances of a master and a considerable number of advanced research students working together over a period of time in a specialty until they obtained, by sheer concentration of effort, an edge over everyone else in the world. Toward the end of the century the laboratories of some of the professors became so famous that the ablest students from all over the world went there for varying periods of time. The list of students who worked in such places often included practically all the important scientists of the next generation. There were few significant physiologists anywhere around 1900 who had not been students of Carl Ludwig at Leipzig. The same was true for psychologists for whom it was a must, in the 1880s, to study with Wilhelm Wundt, who was also at Leipzig.

These unplanned and unexpected developments were an even more de-

---

[26] *Ibid.*, and J. Ben-David, "Scientific Productivity and Academic Organization in Nineteenth Century Medicine," *American Sociological Review* (December 1960), 25:828–843.

german scientific hegemony and the emergence of organized science

cisive step in the organization of science than the early nineteenth century reforms. Research started to become a regular career, and scientists in a number of fields started to develop into much more closely knit networks than ever before. Their *nuclei* were now university laboratories training large numbers of advanced students, thus establishing between them personal relationships, highly effective means of personal communication, and the beginnings of deliberately concentrated and coordinated research efforts in a selected problem area.

The University Outgrows Its Original Functions

Thus emerged the role of professional researcher and the social structure of the research laboratory in the German universities between 1825 and 1870. Their emergence was not the result of any demand for scientific services outside the system of the universities, but of developments within the system itself, which was virtually independent from the other sectors of the society. Experimental science did not have to prove its value for any practical purposes in order to achieve this success. It only had to show its superiority as a method of creating valid new knowledge in universities originally established for a philosophical purpose. However, because these were part of a competitive system, it was inevitable that reward was distributed, above all, according to competence and intellectual achievement measurable by universalistic standards. Thus experimental science gained an upper hand in the universities and maintained this position irrespective of the general cultural and political climate of the society.

There was also very little planning of the process. The role of the professor-researcher was deliberately created by the reformers of the German university, but the original conception had not been that of a laboratory head who directed the work of several researchers but that of a philosopher-scholar who worked on his own and communicated the results of his research to various audiences. The university was to be the place where a few dozen such professors would teach a few hundred selected students about the intellectual foundations of the learned professions. The professors would introduce others into one of the few then accepted scholarly and scientific disciplines to the point where they could teach the subject to high school students, or, if they were able and inclined to do so, continue on their own and eventually become researchers. Unexpectedly, however, in the empirical sciences, research organizations arose that required growing investment and produced new knowledge of a kind that could not be related any longer to these original purposes.[27]

To give an example, in 1820 chemistry could be taught by a single professor doing his research in his own private laboratory, alone or with the aid of a servant or an assistant. What he could teach, including his own discoveries,

---

[27] About the unplanned and unexpected nature of this development, see Schnabel, *op. cit.*, Vol. 2, pp. 209–210, and Vol. 5, pp. 274–275.

was exactly what was needed for the preparation of a senior high school chemistry teacher and not much more than what a bright medical student was likely to be interested in.

By 1890 the field was too complex to be handled even by four professors, and most of their own research was of interest only to other active or intended researchers.

There was also growing specialization in the humanities. More and more historical periods and cultures were investigated and taught. But in the humanities the individual pattern of research could be maintained. A professor of Assyrian language did not expect to have more than one or two students and did not need any assistance to do his own research. Furthermore, the investment required for the establishment of a chair in the field was relatively modest and had little effect on anything else that went on at the university.

The situation was quite different in chemistry (or any other field of experimental natural science). The absence of a specialist, let us say in physical chemistry, reflected on the training given in chemistry in general. The establishment of a chair required considerable investment as well as a commitment on the part of the university to train a certain number of students on an advanced level in a new and specialized field, the demand for which was unknown, but clearly unrelated either to high school teaching or to basic medicine, which were the two scientific occupations originally envisaged by those responsible for the university curriculum.

The university outgrew the tasks assigned to it at the beginning of the nineteenth century, and there was an obvious need for a redefinition of its functions and of the roles of the researcher.

Beginnings of Applied Science

The redefinition of the functions that became necessary was a problem which concerned not only the universities but also the place of science in German society in general. Science in Germany grew up as part of a philosophical-educational enterprise and without the support of a powerful scientistic movement. By about 1870, however, the developments in science that have been described here, and the economic and political developments that set Germany on the course of industrialization and emergence of a more equalitarian class structure made science relevant to technology as well as to economic, political, and social problems. Thus science had reached the limits from which it could develop further as a subsystem insulated from the rest of society.

Starting in the decade around 1860 to 1870, the rise of organized laboratory research and the availability of trained researchers made possible a new kind of applied work. An original idea with practical implications could now be explored and exploited within a short period of time by a group working in concentrated fashion. There were two striking examples of this kind: the

126

development of aniline dyes and of immunizing vaccines.[28] Both of these instances led to the establishment of nonteaching research laboratories employing professional researchers who were not professors.

Another development occurred in the institutes of technology. These had no university status in Germany, although the *Eidgenoessische Polytechnik* in Zürich (which was part of the German language academic system) had such a status and was considered a most distinguished institution. In any case, industrial research and technological institutes became increasingly important "consumers" and eventually also producers of university-level science. Thus, although the instances where a scientific discovery became the immediate source of a useful invention remained rare exceptions, science came into close relationship to technology through the scientific training of engineers and increasingly frequent recourse to scientific consultation and research by industry, hospitals, and the military.

The results of rapid scientific growth in Germany were, therefore, similar to what had occurred in Britain and France before (in spite of the differences in the circumstances of the onset of the growth). However, because these results were not consistent in the German case with the declared functions of the universities (which were purely philosophical and scientific), nor with the place of scientists in German society (who were not part and parcel of an upper middle class scientistic movement consisting of businessmen, politicians, and intellectuals), there was a problem of how to accommodate the changes within and without the university.

Rise of the Social Sciences

A closely parallel development was the rise of social sciences. In the absence of an important scientistic movement, there was originally little social thought of a practical kind in Germany as compared with Britain and France. During the second half of the nineteenth century, however, experimental psychology, historical sociology, economics, and in some parts of the system even mathematical economics on a high level started to emerge.

As in the natural sciences, these developments were in no way a response to outside demand, but grew out of purely academic concerns. This becomes obvious if one compares the rise of these fields in Germany with their earlier rise in Britain and France. Experimental psychology in Germany, like the previous British and French speculations about mental phenomena in psychological terms, was an attempt to understand human behavior scientifically. This attempt was an immanently intellectual response to the development of experimental science. If all natural events could be explained scientifically, then human

---

[28] On aniline dyes see D. S. L. Cardwell, *The Organization of Science in England* (London: Heinemann, 1957), pp. 134–137, 186–187; and David S. Landes, "Technological Change and Development in Western Europe, 1750–1914," *The Cambridge Economic History*, Vol. VI, Part 1 (Cambridge: Cambridge University Press, 1966), pp. 501–504.

behavior could not be an exception either. From this point of view the German attempts were but a further link in a chain that started with Descartes and Locke. But whereas in the Western countries the scientistic psychology was a step toward attempts at creating a secular moral philosophy, in Germany the purpose of the psychologists was to revolutionize philosophy as an academic discipline and to obtain academic recognition for their new approach to mental phenomena.[29]

Sociology and economics also related to the academic situation rather than to practical economic and political problems. This relationship is manifested in the overwhelming concern of these disciplines in Germany with historical rather than with contemporary problems (as was the case in Britain and France). German sociologists and economists were not members of a politically active upper middle class, but of a closed academic community. Thus instead of using scientistic concepts to devise models for a liberal and economically progressive society, they tried to create a new methodology for historiography and the other humanities. When Max Weber wanted to understand the peculiar characteristics of modern society, he tried to penetrate the spirit of seventeenth century Puritanism, which he considered as the root of capitalism. The French sociologist, Emile Durkheim, tried to do the same through a theoretical discussion of different types of division of labor and the analysis of suicide rates in the different societies of his own time. In England grandiose social thought in the style of Comte, Marx, and Herbert Spencer did not lead to the rise of sociology as an academic field. Instead, social investigations were conducted by people interested in social reform, such as Charles Booth, Beatrice Webb, and others.[30]

The difference is even more conspicuous in economics. The British-French school was concerned with economic analysis, while the German school was overwhelmingly historical.[31]

Although these academic beginnings were little concerned with applications, they were parallelled by a growing popular interest in these matters. Germany became a partial parliamentary democracy. It was faced with all the problems of how to grasp and manage the affairs of a modern society. As a result there arose ideologies (Marxism), investigations about social problems, and, in psychoanalysis, a heroic effort toward the creation of a science-based morality.[32]

So in these areas as well, scientific development approached practical concerns. The fact that this occurred without the background of a scientistic

[29] Joseph Ben-David and Randall Collins, "The Origins of Psychology," *American Sociological Review* (August 1966), 3:451–465.
[30] Max Weber, *The Protestant Ethic and the Spirit of Capitalism* (London: Allen & Unwin, 1930); Emile Durkheim, *The Division of Labor in Society* (Glencoe, Ill.: The Free Press, 1947); *Suicide: A Sociological Study* (Glencoe, Ill.: The Free Press, 1951); Beatrice Webb, *My Apprenticeship*, 2 vols. (Harmondsworth: Pelican Books, 1938).
[31] See footnote 4.
[32] Philip Rieff, *Freud, The Mind of a Moralist* (New York: Viking Press, 1959).

german scientific hegemony and the emergence of organized science

movement based on broad social and political support created in the field of social sciences an even more acute problem than in technology.

Early Twentieth Century Role of the University in German Society

Our discussion will now consider two questions: (1) How did the university respond to the changes that occurred within it, namely the potentially very large expansion of the scope of the fields investigated and taught at an advanced scientific level and the transformation of research in several fields into an increasingly large-scale organized operation? And (2), to what extent was the relationship of the university to its environment modified as a result of its growing relevance to technology and contemporary affairs? The discussion will attempt to deal with the first question systematically and to comment on the second only in a general way.

Quantitatively, the expansion of the university and of its research activities was very rapid. The number of university students doubled between 1876 and 1892 from 16,124 to 32,834; by 1908 it rose to 46,632. In the institutes of technology, which were given university status in 1899, the numbers rose from 4,000 in 1891 to 10,500 in 1899. The growth of the academic staff was somewhat slower, but it started earlier (1,313 in 1860; 1,521 in 1870; 1,839 in 1880; 2,275 in 1892; 2,667 in 1900; and 3,090 in 1909). The total university budgets of Prussia, Saxonia, Bavaria, and Württemberg were 2,290,000 Mark in 1850; 2,961,000 in 1860; 4,734,000 in 1870; 12,076,000 in 1880; 22,985,000 in 1900; and 39,622,000 in 1914.[33]

But at the same time there was a building up of tensions within the university. Rather than change their structure so as to be able to take full advantage of the expanding opportunities, the universities adopted a deflationary policy of restricting the growth of new fields and the differentiation of old ones. Although the number of students and staff increased, and although there was an even greater increase in the expenditure of the universities because of the steeply growing expense of research, no modifications were made in the organization of the university. Officially, it remained a corporation of professors even though their ratio to other academic ranks, which included extraordinary

---

[33] W. Lexis (ed.), *Die deutschen Universitäten: für die Universitätsaustellung in Chicago, op. cit.* See Vol. I, pp. 119 and 146, and for other countries, p. 116; W. Lexis (ed.), *Das Unterrichtswesen im deutschen Reich*, Vol. I, *Die Universitäten* (Berlin: A. A. Ascher, 1904), pp. 652–653; and Friedrich Paulsen, *Geschichte des gelehrten Unterrischts an den deutschen Schulen und Universitäten vom Ausgang des Mittelalters bis zur Gegenwart*, 3rd ed. (Berlin and Leipzig: Vereinigung Wissenschaftlicher Verleger, 1921), Vol. II, pp. 696–697. For scientific developments see Lexis, *Das Unterrichtswesen*, pp. 250–252; and Frank Pfetsch, *Beitraege zur Entwicklung der Wissenschaftspolitik in Deutschland*, Forschungsbericht (Vorläufige Fassung), (Heidelberg: Institut für Systemforschung, 1969), (stencil), Part B, Appendix, Table IV.

professors and *Privatdozenten,* who had some academic standing, and institute assistants (who had no formal academic standing at all), changed considerably. This organization was particularly visible in the experimental natural sciences and in the social sciences, which had the greatest potentiality for growth. In the case of the natural sciences, the development was probably due to the growth of research institutes that encouraged professors in experimental science to regard their respective fields as personal domains. The growth of the social sciences was prevented mainly by the difficulty of keeping political controversy apart from empirical inquiry in those ideologically sensitive fields. All this led to a sense of frustration and hopelessness in the academic career manifested by the rise of organizations resembling trade unions in the lower ranks. The *Vereinigung ausserordentilicher Professoren* was founded in 1909; the *Verband deutscher Privatdozenten* in 1910; and two years later the two organizations were fused into the *Kartell deutscher Nichtordinarier.*[34]

The difficulty experienced by the aspiring scientists and scholars was largely a consequence of the conservatism of the university organization and the professorial oligarchy which dominated it. The professors, who as a corporate body *were* the university, prevented any important modification of the structure that separated the "institute," where research took place, from the "chair," the incumbent of which was a member of the university corporation. The former was like a feudal fief of the latter. The result of this system was that while the increase in research activity fostered an unbroken progression from the mere beginner to the most experienced and successful leader in a field, the organization of the university obstructed this progression as a result of the gap in power and status between the professor who had a chair and all the others who did not.

A closely related manifestation of this conservatism of the highly privileged university faculties was their resistance to any innovation of a practical or applied character. Not only did they not admit engineering studies to the university, but they also resisted the granting of academic degree-awarding powers to institutes of technology (the right was nontheless conferred by the emperor in 1899). They also resisted the recognition of the *Realgymnasium* as a qualifying preparation for the universities and many other proposals for

---

[34] For the ratio of full professors to other ranks, see Lexis, *Die deutschen Universitäten,* p. 146, and *Das Unterrichtswesen,* p. 653. For the differences between fields, see Christian von Ferber, "Die Entwicklung des Lehrkörpers der deutschen Universitäten und Hochschulen, 1864–1954," in H. Plessner (ed.), *Untersuchungen zur Lage der deutschen Hochschullehrer* (Göttingen: Vandenhoeck und Ruprecht, 1956), pp. 54–61 and 81. For the foundation of the various unions, see Paulsen, *op. cit.,* p. 708. The whole problem has been surveyed in Alexander Busch, *op. cit.,* and "The Vicissitudes of the *Privatdozent:* Breakdown and Adaptation in the Recruitment of the German University Teacher," *Minerva,* I (Spring 1963), I:319–341. The difficulties in the social sciences are described in Anthony Oberschall, *Empirical Social Research in Germany 1868–1914* (Paris and The Hague: Mouton, 1965), pp. 1–15 and 137–145.

german scientific hegemony and the emergence of organized science

reform. Their resistance to the study of bacteriology and psychoanalysis has been described elsewhere.[35]

In the basic fields and earlier established fields with direct applications, expansion continued but became selective. Among the existing and well-established fields there was a rapid growth of new university chairs only in mathematics and physics. In the other well-established fields, there was little expansion.[36] Intellectually important innovations, such as physical chemistry, physiological chemistry, and other fields were only grudgingly granted academic recognition.[37] Specialists in these fields received titles of *Extraordinarius*, or institute head, but they were only very sparingly given the rank of *Ordinarius*—the only "real" professors—and then not through the establishment of new chairs but by the appointment of individuals to existing chairs with loosely defined terms of reference. Mostly, the great growth of research and specialization led only to a swelling of the ranks of assistants. Sociology, political science, and economics were only rudimentarily developed as independent fields. The main developments took place through the establishment of new chairs in clinical medicine and in the increasing number of languages, literatures, and histories taught in the humanistic faculties.[38]

This pattern of expansion shows that the competitive mechanism which had previously ensured the prevalence of purely scientific considerations in the establishment of new fields of study was impaired. In theoretical fields that did not require large-scale laboratory facilities (theoretical physics, mathematics, theology, humanities) it worked as before. Where laboratory facilities were required, however, rapid growth occurred only in experimental physics and clinical medicine.[39] Both of these were new fields which did not compete with any established discipline (important laboratories in physics emerged only in the 70s, and clinical research in specialized fields also started only at that time).[40] Their growth might also have been stimulated by the external competition of the institutes of technology, new governmental research institutes, and public

[35] Joseph Ben-David, "Roles and Innovation in Medicine," *American Journal of Sociology* (May 1960), LXV:557–568.

[36] See von Ferber, *op. cit.*, pp. 71–72, and Zloczower, *op. cit.*, pp. 101–125.

[37] *Ibid.*, pp. 114–115 (about physiological chemistry). Even in such a theoretically important field as physical chemistry, there were only five institutes—Leipzig, Berlin, Giessen, Göttingen, and Freiburg—and five subordinate positions (*Extraordinarii* who had special departments in institutes headed by someone else) in Breslau, Bonn, Heidelberg, Kiel, and Marburg in 1903. This was the case almost 20 years after Ostwald founded the first chair in the field in Leipzig (1887) and more than 20 years after the publication of his famous textbook and the launching of a journal in the field (see Lexis, *Das Unterrichtswesen*, pp. 271–273).

[38] von Ferber, *op. cit.*, pp. 54–61. For the whole problem see L. Burchardt, "Wissenschaftspolitik und Reformdiskussion im Wilhelminischen Deutschland," *Konstanzer Blätter f. Hochschufragen* (May, 1970), Vol. VIII: 2, pp. 71–84.

[39] *Ibid.*, pp. 71–72, and A. Zloczower, *op. cit.*, pp. 101–125.

[40] *Ibid.*, and Felix Klein, "Mathematik, Physik, Astronomie," in Lexis (ed.), *Das Unterrichtswesen*, Vol. I, pp. 250–251.

131

hospitals with good research facilities. Even here development was held back. University physics laboratories were considered inadequate by the 1880s, and clinical fields were considered as substantive specializations within the established basic medical discipline. These specializations were not given the autonomy accorded to the new clinical research units that started to emerge in the United States toward the end of the century.[41]

This new situation was a reversal of the trend that brought the experimental sciences to the forefront of the universities. The change was not the result of the intellectual exhaustion of those disciplines, but, as has been pointed out, of the impairment of the competitive mechanism. In other countries (e.g. the United States) personnel in the experimental sciences continued to grow at a faster rate than that of other disciplines, and even in Germany the number of students in the natural sciences grew at twice the rate of the growth of students in the philosophical faculty in general between 1870 and 1912.[42] The effectiveness of competition had been dependent on the opportunities for innovators of new specialities (usually young) to establish their names independently of their teachers and to obtain a separate new chair and laboratory. They used to obtain these first at a peripheral university, and then success would compel the other universities to follow suit. With the growth of institutes in the older experimental fields (chemistry, physiology), it became impossible for a young person to establish himself without the aid of a professor, because no one could do significant research outside the institutes any longer. This development increased the power of the heads of the institutes, who had vested interests in keeping new specialities that arose in their fields as subspecialties within their own institutes rather than allowing them to become separate chairs with claims for new institutes.

Thus, while a kind of class tension was built up within the established fields, there was increasing resistance to the institutional provision for the cultivation of new fields. If the latter were innovations of an intradisciplinary or pure scientific character, they were usually accommodated within the university, but often in a way which made them subordinate to older disciplines. There were obstinate and long, drawn-out debates about the theoretical importance of new fields to justify the establishment of any new chair. These debates, which were often conducted in terms of personal qualifications of particular candidates, often obscured the real issues and also introduced a great deal of personal bitterness into academic matters. Had the universities been organized as departments, these matters might have been treated in quite an impersonal fashion.

Thus the rigidity of the corporate organization came into full play. As long as the social unit of research was the individual and the fields were few and distinct, the system of independent chairs suited the requirements of

[41] Pfetsch, *op. cit.*, Part B, pp. 27–32; Felix Klein, *op. cit.*, pp. 250–252; A. Flexner, *Medical Education: A Comparative Study* (New York: Macmillan, 1925), pp. 221–225.
[42] Pfetsch, *op. cit.*, Part B, p. 35.

german scientific hegemony and the emergence of organized science

research reasonably well. Disciplinary innovations did not require changes in the organization; they required only the addition of a new chair to the existing ones. Competition could compel the universities to do this.

But once the basic unit of research became the group, and the boundaries of the fields became increasingly blurred, there was a need to change the organization. The corporate university was unwilling to change, however, and, in the absence of a strong university administration, competition was not enough to force it to do so. The advantages of attracting a good man to the university could persuade its members to consent to the establishment of a new chair, especially if the problem arose when the person with a possible vested interest in the matter just vacated his chair. But to try to persuade several colleagues in related fields to enter into a cooperative relationship with each other, or take a stand in potential jurisdictional disputes between powerful institutes (which were considered the private domain of professors), or in conflicts between an institute head and his assistants, was something that no corporation of equals would undertake. They preferred the establishment of new nonuniversity research institutes to making changes in the structure of the universities. The initiative in the new fields of science was therefore addressed to the central government. Thus arose the *Physikalisch-Technische Reichsanstalt* in 1887, and the *Kaiser Wilhelm Gesellschaft* (now *Max Planck Gesellschaft*) in 1911.[43]

The Place of the University in the Social Structure of Germany

The discussion now proceeds to the second question: to what extent was the relationship of the university to its environment modified as a result of its growing relevance to practical matters? In contrast to the conservativeness of the universities, the governments of the main German states as late as 1933 seemed to have been relatively generous and forward looking in their support. As has been shown, the institutes of technology were given university status by imperial decree. The *Kaiser Wilhelm Gesellschaft*, which established some of the foremost research institutes in Germany, was similarly supported by the government, and the university budgets themselves were rapidly increased. Even private industry provided important support for research. From all the accounts of the situation during the period preceding World War I

[43] Busch, *op. cit.*, pp. 63–69. The share of the universities in the total expenditure on science declined from 53.1 percent in 1850 to 40.4 percent in 1914, while those of nonuniversity research institutes in the natural sciences (not including medicine and agriculture) rose from 1.4 to 11.0 percent, and those of the institutes of technology, from 5.3 to 13.4 percent. Since these data also include the humanities, in which the share of the universities increased during the period, the shift of science away from the universities was probably even greater than that indicated by these figures. See Pfetsch, *op. cit.*, Part B, Table IV, summarizing expenditures of the Prussian, Bavarian, Saxonian, Baden and Württemberg states. This shift confirms the interpretation that the initiative slipped away from the universities to the specialized research institutes and the institutes of technology.

(and even the period of the Weimar Republic) there emerges a picture of growing and effective support of research.[44]

But this support did not alter the basic problem of the redefinition of the functions of the university, or indeed of science in general, in society. Support was given for science and even for scholarship because they were considered effective means for military, industrial, and diplomatic purposes (for instance the studies of foreign language and cultures).[45] This support was not much different from that in other countries, and, like elsewhere, scientists knew how to use this support for their own purposes. What was different from other countries (especially Britain and the United States, but to some extent also France), was the absence of feedback of values from science to society. *Scientific knowledge* was highly valued and widely diffused as one of the ingredients of technical expertise. But the *values of science* as ingredients of social and economic reform, and as professional ethics were not. The kind of movement which in the West attempted to professionalize an increasing number of occupations and imbue the business classes, technologists, the civil service, politicians, and people in general with the values of scientific universalism and altruism through higher education and social research was missing.[46] Or, to the extent that it existed, it had very little to do with the university or with professional science in general.

The reason for this was as follows: Intellectual inquiry in Germany did not thrive as part and parcel of the way of life of a middle class of independent people whose position was based not on privilege but on achievement in various fields. It started thriving as a hothouse flower mainly on the support of a few members of the ruling class. From the point of view of the intellectuals the universities created under the particularly favorable conditions of the struggle against Napoleon were the only secure institutionalized framework for free

[44] Pfetsch, *op. cit.*, Part B, pp. 4–8, and J. D. Bernal, *The Social Function of Science* (London: Routledge & Kegan Paul Ltd., 1939), pp. 198–201. As a percentage of the gross national product there was no significant growth between 1900 and 1920 and probably not too much after that either.

[45] This may not be apparent from the breakdown of science budgets by stated purpose of expenditure. But the role of military-industrial interests in support of science becomes evident in case histories of the foundation of important research institutes. See Pfetsch, *op. cit.*, Part B, pp. 27–32; Part C, pp. 14, 56–59; and Busch, *op. cit.*, pp. 63, 66–69.

[46] About the connection between the scientist movement, social mission, and professionalism in Britain and the United States, see Webb, *op. cit.*, Vol. I, pp. 174–197; Vol. II, pp. 267–270, 300–308; R. H. Tawney, "The Acquisitive Society" (New York: Harcourt Brace Jovanovich, 1920), Chap. VII; A. Flexner, *Universities: American, English, German* (New York: Oxford University Press, 1930), pp. 29–30; Armytage, *op. cit.*, and N. Annan, *op. cit.* In contrast to this approach, the support of science in Germany was not a subject matter for public, or even parliamentary, debate. Decisions about science were made in a somewhat similar fashion to those made about military matters, namely by governments acting on the advice and/or under the pressure of small closed circles of expert civil servants, scientists, and industrialists. See Pfetsch, *op. cit.*, Part C, pp. 60–61. About the absence of educational ideals and practical preparation of would-be professionals in Germany, see Wende, *op. cit.*, pp. 126–127.

intellectual activity in the country. The status and the privileges of the universities were granted to them by the military-aristocratic ruling class, and were not achieved as part of the growth of free enterprise. It was, therefore, a precarious status based on a compromise whereby the rulers of the state regarded the universities and their personnel as means for the training of certain types of professionals. However, the rulers allowed the universities to instruct in their own way and to use their position for the pursuit of pure scholarship and science. Therefore the universities had to be constantly on the defensive, lest they be suspected of subversion and lose the elite position which ensured their freedom.[47]

The rise of the middle classes and of an industrial working class toward the end of the nineteenth century provided an opportunity to change this outlook. The universities, which by this time had gained very great prestige, might have played an important role in helping these new socially advancing elements to develop an equalitarian and universalistic social ethos such as had developed in the West in the scientistic liberal and in the later socialistic movements. Instead, however, the universities, like the other privileged segments of society, chose either to oppose these new developments, or, at the very least, to stand apart from them.

Besides the fear that certain intellectuals have always and everywhere tended to entertain about the debasement of standards as a result of growing equalitarianism, there was an additional problem in Germany. Scientists there had little incentive to become part of the bourgeois middle classes, since these had little official dignity and even less self-dignity. The purpose of the German bourgeois was to be absorbed into the aristocracy.[48] A highly respectable bourgeois stratum comprising businessmen and professionals did not exist in Germany as it did in France. Nor were there successful and powerful upper and

47 See Ben-David, *op. cit.*, and Zloczower, *op. cit.* This is a crucial point in the understanding of German universities. Ostensibly, they were founded with definite moral aims in mind. Fichte, Humboldt, and others believed that the university would create a philosophically trained and morally upright leadership group, and as far as the elite of the civil service and academic life was concerned, this was actually the case to some extent. But Fichte and Humboldt did not consider the possibility that conflicts might arise between the state and the university about practical political matters. They viewed the state as a representative of their own idea of culture (*Bildungsstaat*), and between the state and the university they felt there could be only a division of labor but no conflict of interest about fundamental issues. The preestablished harmony between the state and the university envisaged by this view could be maintained, however, only as long as (a) the government was above all party struggle and clash of economic interests in the country and (b) the university also remained completely insulated from social interests and conflicts. When the new German university was founded, the Prussian "state" seemed to be in such a position of independence from "society." When this view became untenable, the university was faced with the choice of becoming part of the ordinary civic society, as it was in France and England. However, it never accomplished that aim. About the situation early in the nineteenth century, see König, *op cit.* About the continuous malaise felt in face of the developments before and after World War I, see Fritz K. Ringer, *The Decline of the German Mandarins: The German Academic Community, 1890–1933.* For a contemporary view, see Schelsky, *op. cit.*, pp. 131–134.

48 Sombart, *op. cit.*, pp. 448–450.

german scientific hegemony and the emergence of organized science

upper middle class groups as in Britain—groups that shared the values of science and were able and willing to support the case of science politically and economically.

The only class that had a comparable dignity and somewhat comparable outlook and interests to those of the professors was that of the higher civil servants. These persons were trained at the universities, were often highly educated, and were imbued with a sense of mission. They accepted the idea of the *Kulturstaat*, which made it incumbent on them to support higher learning and creativity because they were educated in this spirit and because this was a useful legitimation of their claim for quasi-aristocratic status and authority over all the other classes of society.

This constellation of social classes explains the behavior of German academics toward the new social developments. The choice appeared to them as one between siding with a public-spirited and educated aristocracy of merit, and a selfish, poorly educated middle class that could not even be trusted with the protection of business interests, as its members were always willing to join the reactionary landed aristocracy at the first opportunity. The working class did not present a respectable alternative before World War I. Afterwards it could have provided such an alternative, but the majority of professors did not consider it as an attractive or even a possible alternative, as most of them identified socialism with mob politics and anticreative equalitarianism.[49]

There was nothing particular to Germany in these attitudes. They were shared by many academics in the West and elsewhere. Yet in Britain and the United States even some of the academics sharing these conservative aristocratic attitudes drew from them the conclusion that it was their duty to try and make these new classes more cultured as well as more public-spirited. In France, conservatives, like people of other political persuasion, might have adopted any course that seemed reasonable to them individually, irrespective of what others did, and they could have carried their feud with their colleagues holding different views to wherever they taught or acted in whatever capacity. In Germany, however, an attempt was made to disregard the whole problem by stressing the university's neutrality. This position made it possible for the liberals to accept the privileged, quasi-aristocratic standing of the university, to justify its steeply hierarchic internal structure, and to rationalize its aloofness from issues that might involve value judgments and raise passions.[50] Under the circumstances, however, the uninvolvement of the university in politics, contemporary affairs, and even technology could be easily abused by conservative and generally right-wing interests. Due to the traditional identification of the university and the idealized state and the close relationship of interests between the higher civil

---

[49] Ringer, *op. cit.*, pp. 128–143.

[50] Max Weber, "The Meaning of 'Ethical Neutrality' in Sociology and Economics," in his *On the Methodology of the Social Sciences*, tr. and ed. by Edward A. Shils and Henry A. Finch (Glencoe, Ill.: The Free Press, 1949), pp. 1–47.

servants and the professors, anything that had to do with the state was at least partially exempt from the rule of noninvolvement.

The universities, which were extremely wary about the recognition of social sciences and anything that had to do with contemporary affairs, nevertheless tolerated the abuse of the study of history and literature for nationalist or anti-Semitic propaganda.[51] They did not find the open and official involvement of many of their members with military affairs inconsistent with their opposition to the introduction of technological studies at the universities.[52] Thus there was a great deal of ambiguity about the neutrality of the universities since their faculties allowed the universities to be used for the political purposes of a more or less absolutistic government and as a forum by those appointed as the true representatives of the traditional order.

It is difficult to determine with any certainty the reason that so many of the "liberal" elements accepted this situation. Prior to World War I they could justify their behavior by the identification of the Wilhelmian Empire with the idea of the *Bildungsstaat,* or even by regarding Germany as the most socially advanced country of the world. The refusal of the universities to cooperate with the attempts at university reform initiated by Becker in the Weimar Republic, the nostalgia of so many of the professors to the old order, and their delegitimation of the new one is more difficult to explain in the same terms. The academic attitudes at both points of time were, however, consistent with the interest of the professors to maintain themeslves as a highly privileged "estate" elevated above the classes of society, not accountable to any, but protected by and spontaneously allied with a similarly privileged higher civil service. It seemed inconvenient to exchange this situation with one where the university would have to deal with all kinds of politicians.[53]

Because of the internal tensions between the ranks and the difficulty of obtaining recognition for new fields, the center of scientific activity, especially in some of the latter fields, started to shift to Britain and the United States.[54] The political tension which arose after that war, and the chronic problem of graduate unemployment made the position of the university in society increasingly difficult. If Germany nevertheless maintained scientific leadership, this was due partly to the existence of a very large group of scientific leaders who

[51] R. H. Samuel and R. Hinton Thomas, *Education and Society in Modern Germany* (London: Routledge and Kegan Paul, 1949), pp. 116–118, and Peter Gay, "Weimar Culture: The Outsider as Insider," in Donald Fleming and Bernard Bailyn (eds.), *Perspectives in American History* (Cambridge, Mass.: Charles Warren Center for Studies in American History, Harvard University Press, 1968), pp. 47–69.

[52] Busch, *op. cit.,* p. 63. About the discrimination against technology, see Wende, *op. cit.,* p. 133.

[53] Schelsky, *op. cit.,* pp. 164–171.

[54] On early U.S. superiority in astronomy, cytology, genetics, certain branches of physics, medicine, industrial research, and animal behavior, see J. D. Bernal, *The Social Functions of Science* (London: Routledge and Kegan Paul, 1940), p. 205. On the shift in the center of research in physiological chemistry to Britain, see Zloczower, *op. cit.,* p. 115.

grew up before the war, and partly to the inertia of the international scientific community, which continued to use the German universities as its favorite training ground and meeting place. Uninvolved in the political tensions and occupational uncertainties of their hosts, the visiting scientists saw the German university in the light of its ideal self: as a seat of the purest learning for its own sake and an unequalled center of overall excellence.[55] Under these conditions it was easy to maintain German scientific supremacy by judicious governmental intervention in the affairs of science. This, situation, however, could last only as long as there was a government interested in the maintenance of such supremacy; the university system ceased to be a source of scientific initiative and drive, and there was no other social mechanism (except the wish of the government) to replace it.

It is a futile question to ask whether the shift could have been reversed, if the Nazis had not taken over the country, as the universities were part of the system which made the Nazi take-over possible.

[55] Charles Weiner, "A New Site for the Seminar; The Refugees and American Physics in the Thirties," in Donald Fleming and Bernard Bailyn (eds.), *Perspectives in American History*, Vol. 2 (Cambridge: Charles Warren Center for Studies in American History, Harvard University, 1968), pp. 190–223.

the
professionalization
of research
in the united states
eight

The Graduate School
in the United States

The changes that occurred in the United States between the 1860s and the time of World War I consisted in some cases, of logical conclusions to developments which had started in Germany. This was the case in the development of the graduate school and the organization of university research. However, in the training given for the professions, and to an even greater extent in the program for undergraduate education, the German influence was adapted to more indigenous American or rather to a common American-British tradition.

The crucial step in the importation of the European model was the establishment of the graduate school. Although, properly speaking, there was no graduate school in Germany—and it still exists only in rudimentary form—those who initiated the graduate school in the United States believed that they were closely following the Germany model.[1]

German and other European universities trained their students for a single-level degree. When the system was established early in the nineteenth

[1] Lawrence R. Veysey, *The Emergence of the American University* (Chicago: University of Chicago Press, 1965), pp. 160–161 and 166. The desire to follow the German model as closely as possible was especially marked in the faculties of the graduate schools. University presidents tended to be more pragmatic.

century, it was in fact possible to give a complete and well-rounded training in any branch of science and scholarship at this level. After all, many outstanding scientists were still amateurs, and a single professor could invariably master a whole field. The philosophical faculty of the German university (which included all the humanistic and scientific subjects) provided a scientific or scholarly education up to the highest standards. But not all who obtained the degree were qualified to do research. The conception of a professionally qualified research worker existed nowhere early in the nineteenth century, because research was considered a charismatic activity that could be successfully pursued only by an inspired few. The university could, however, and the Germany university did in fact, make a serious attempt to teach at the highest level everything that could be taught in the major academic disciplines.

By the end of the century, however, the single-degree program became an anachronism. The university still pretended that its degree course was at the highest scientific level, and some of the teaching more or less conformed to this ideal. But even in these cases it was impossible to obtain a training for independent research within the confines of such a program. Those who were to become research workers acquired their specialized knowledge and skills informally as assistants working with professors in the research institutes, usually attached to the chairs, where they had the benefits of doing serious research and of contact with a number of more advanced assistants. The uncompromising level of the degree course was more than the student who did not intend to enter research could usefully assimilate, yet it was not enough for those who wished to enter a professional research career. The training of the latter remained informal. Its main shortcoming was that it was difficult for the student to acquire an all-round training in his field, because he worked with a single teacher. This system also created a situation of dependence on a teacher who often behaved in a highly arbitrary and authoritarian manner, and it gave rise to feelings of insecurity among those who aspired to a research career. As long as the student was not appointed to a university chair, he remained an assistant in a bureau-cratic framework with little independent professional standing, even if he was an advanced research worker performing important tasks in research as well as in the training of beginners.[2]

For the American and British (and perhaps other foreign) students who went to Germany, all these shortcomings were not obvious. They were an already highly selected group who possessed a first degree and occasionally even some research experience. The problems of the academic career in Germany did not disturb them, since their own careers were not dependent on the German pro-fessor. The fact that the training was not adapted to the needs of the German student who had to acquire well-rounded skills made it all the more appropriate

[2] See A. Zloczower, *Career Opportunities and the Growth of Scientific Discovery in 19th Century Germany* (Jerusalem: The Hebrew University, The Eliezer Kaplan School of Economics and Social Sciences, 1966), pp. 64–66.

140

the professionalization of research in the united states

to the needs of visiting graduate students, who often had clear-cut ideas about what to study and with whom. Nor, apparently, were they acutely aware of the problems that arose from the bureaucratic subordination of the assistant to the heads of the institutes. As welcomed visitors they did not have any difficulty in being admitted to institutes or in moving from one institute to another. From their point of view the institutes were integral parts of the university where research and training for research were performed.[3]

One of the results of this misconception was that when American or British scholars returned to their countries advocating the adoption of the German pattern, they did not make any distinction between the chair and the institute. Although they knew that the German professors personally acted in a very hierarchic manner, they were unaware of the structural counterpart. They did not see how different the departmental structure was from the combination of chair and institute that they admired and thought they were establishing in their own universities. Nonetheless, the departmental structure eliminated the anomaly whereby a single professor represented a whole field, while all the specializations within that field were practiced only by members of research institutes who were merely assistants to the professor.

What the American pioneers in the establishment of graduate schools had in mind were students such as they themselves had been in Germany—those possessing a first degree and committed to a professional career in research. In Germany, research was not recognized as a profession; it was a sacred calling or vocation for very few persons, who needed no formal training beyond what was offered for the standard degree courses. There was no conception of a career leading to the top by gradual steps. The highest positions were rewards for exceptional accomplishment rather than the orderly culmination of a career. In the United States there was from the very beginning an important innovation in the way the idea of the university as a research-based teaching institution was conceived. The idea that research and teaching at the graduate university could not be determined by anything except the state of science and the creativity of the professors was put into effect much more radically in the United States than in Germany. As a consequence of the uncompromising "idealism" implicit in this view, a far better organized system had developed for training professional research workers. In Germany all students of science or humanistic subjects were required to study their subjects in a highly specialized way, not in order to use them later in life (except for a small minority who followed academic and scientific careers), but because this was considered good for them by those who

[3] Interesting accounts of the experiences of American students in Germany can be found in Ralph Barton Perry, *The Thought and Character of William James*, Vol. I (Boston: Little, Brown and Company, 1935), pp. 249–283, and Donald Fleming, *William H. Welch and the Rise of Modern Medicine* (Boston: Little, Brown and Company, 1954), pp. 32–54, 100–105, and Samuel Reznick, "The European Education of an American Chemist and Its Influence in 19th Century America: Eben Norton Horsford," *Technology and Culture* (July, 1970), XI:3, pp. 366–388.

were in authority. In the United States only the graduate student in arts and sciences was invited to pursue science or scholarship for its own sake, and for him this was a preparation for a career of research. If he did not want to become a research worker, he could limit his education to the traditional undergraduate college or else he could go to a professional school. The graduate school could therefore concentrate on the training of research workers.

The Professional School

The professional school was another structure that enabled the American university to avoid the intellectually constricting influence of the Germanic professorial system. In its undergraduate form, the professional school in America started as a pragmatic experiment in the land grant colleges during the 1860s.[4] But at the postgraduate level, its development was parallel to that of the graduate school in the arts and sciences. To some extent it was also an outgrowth of trends inherent in the state of science around 1900.

According to the conception prevailing in the German universities during the first half of the nineteenth century, the basic scientific and humanistic disciplines had a monopoly of higher education. These disciplines were also emphasized in the training of physicians, lawyers, and clergymen. The monopoly was based on the assumption that teaching at university level had to be creative and based on original research; educators also believed that serious research existed only in the basic scientific and humanistic disciplines. This approach was usually less than optimal for the training of students for the practical professions. Even admirers of the German system admitted that the clinical training of the British doctor was superior to that of his German counterpart. But the emphasis on the basic medical fields was justified by the not unreasonable argument that the practical side of medicine could be acquired in apprenticeship outside the university.[5]

During the second half of the nineteenth century, however, a new kind of research arose that invalidated the assumption that creative research existed only in the basic fields. The discovery of the bacterial causation of illness, the growing amount of engineering research (especially in electricity), the development of psychoanalysis and in a way all social science research, were not "basic"

---

[4] For an exposition of the tradition of the land grant colleges, see James Lewis Morrill, *The Ongoing State University* (Minneapolis: The University of Minnesota Press, 1960); and for an evaluation, see Mary Jean Bowman, "The Land Grant Colleges and Universities in Human Resource Development," *Journal of Economic History* (December 1962), XII:547–554.

[5] The best exposition and the most convincing attempt at the justification of the German university is to be found in the different writings of Abraham Flexner, namely *Universities: American, English, German* (New York: Oxford University Press, 1930), and *I Remember* (New York: Simon & Schuster, 1940). For an exposition of some of the shortcomings see Friedrich Paulsen, *Geschichte des gelehrten Unterrichts an den deutschen Schulen und Universitäten vom Ausgang des Mittelalters bis zur Gegenwart*, 3rd ed. (Berlin and Leipzig: Vereinigung Wissenschaftlicher Verleger, 1921), Vol. II, pp. 710–738.

142

in the accepted sense of the word. The questions asked by inquirers in these fields did not derive from the state of any given discipline. For instance, for the professional physiologist and pathologist seeking to understand bodily functions in physical and chemical terms, the statistical inquiry of Ignaz Semmelweiss into the etiology of puerperal fever made no theoretical sense. And initially the same applied to the discovery by Pasteur and others of the bacterial causation of illness.[6] From the point of view of "normal, puzzle-solving" science, these investigators asked the wrong questions and got meaningless answers. The fact that some of these answers had dramatic practical uses made matters even more disturbing.

However, this kind of research grew into a regular activity. It assumed the characteristics of a discipline. There was a permanent exchange of information between groups of research workers, and they agreed on what constituted a problem and what were the proper models of research to solve them. They trained entrants into the field, as basic scientists did, even though the relationship of this inquiry to basic scientific theory was often obscure. What has been called "applied" or "problem-oriented science" had begun, and certain aspects of it acquired the social structure of academic disciplines. The word "quasi-discipline" will be used to distinguish them from those fields that originated from attempts to solve problems defined by the internal traditions of a given science.[7]

With the rise of this quasi-disciplinary research, however, the whole question of the relationship between higher education and professional training was reopened. There was great pressure to make engineering an academic field, and there were similar pressures from other quasi-disciplines.

The attitude toward these developments at the German universities was, with few exceptions, negative. As was pointed out in the previous chapter, the universities preferred to define their task conservatively and to leave this type of research to other institutions.[8]

This approach might have been a satisfactory solution had these other institutions been able to compete with the universities on equal terms, as was the case in physics, mathematics, and perhaps to some extent in chemistry. Institutes of technology, the Kaiser Wilhelm Gesellschaft, and to some extent industry itself, provided in these fields alternative opportunities for research. There were problems here too, however. In chemistry, for instance, applied research laboratories existed in industry, but training took place in the universities. This division retarded the development of chemical engineering as a profession.

[6] Joseph Ben-David, "Roles and Innovations in Medicine," *American Journal of Sociology* (May 1960), LXV:6, 557–568.
[7] This term may be useful to distinguish that kind of applied research which assumes the form of an academic discipline from the kind of applied research which does not. It is impossible to say what this difference is due to, but it is probably related to the intellectual quality of the innovation and the usefulness of training people in the field.
[8] See Chapter Eight, pp. 129–133 and footnotes to those pages.

143

Furthermore, even in the institutes of technology, the acquisition of advanced research skills continued to be dependent on personal apprenticeship. Finally, even in the new, nonacademic research institutions, research was not considered a professional career, so that the research worker who was neither a professor nor a head of an institute had to work in a stiffly hierarchical structure that curtailed his scientific freedom and initiative. But since the opportunities were expanding rapidly in consequence of the recognition of the technological institutes and the foundation of the Kaiser Wilhelm Gesellschaft, these limitations probably did not seriously impair development until World War I.

In the life sciences, which were an exclusive preserve of the universities, the situation was more difficult. As has been pointed out, the universities opposed the development of bacteriology. This too was left to special institutions. The universities also did little for physiological chemistry. Clinical research was developed up to a point by the numerous *Privatdozenten* and *Extraordinarii* who had very strong incentives to stay at the university. Although it offered them no academic career, the university did further their professional medical practice both intellectually and financially. The official structure of the universities, however, took relatively little cognizance of these developments. New chairs were established, but research was dominated by the basic disciplines, and the training of practitioners was influenced little by the new developments. The German universities did not accept the idea that the university had an active role to play in medical practice. Nor did they accept the concept of making practitioners more effective users of research by encouraging research relevant to practice, and by actually training the student in the detailed skills of medical practice in an environment where research constantly tested and modified these skills.[9] Students were still largely taught what was considered the intellectual basis of their profession, and they were expected to acquire the skills needed for either research or practice by their own efforts after graduation. But the relationship between the intellectual basis and the practice that had existed in the first half of the nineteenth century was completely changed by the end of the century, and this change was not sufficiently reflected in the medical faculties.

This attitude was reversed in the United States. There the principle was accepted that universities were training students for the intellectual-practical professions and that this was justified because of their important scientific basis. As a result, even the most research-oriented schools interpreted their task of enriching the scientific element of the professions as an obligation to encourage quasi-disciplinary research relevant to professional work and to train practitioners capable of benefiting from research. The most conspicuous and successful instance was the development of clinical research in medicine at Johns Hopkins University. Instead of emphasizing the invidious difference between basic and clinical re-

---

[9] Abraham Flexner, *Medical Education: A Comparative Study* (New York: Macmillan, 1925), pp. 221–225.

the professionalization of research in the united states

search (even though the theoretical and experimental deficiencies of clinical research were known), attempts were made to create university hospitals with conditions that approximated as nearly as possible the conditions of an experimental laboratory and to use these facilities for the improved training of physicians.

Similar policies were pursued in engineering, agriculture, and education. The relevant departments of the universities considered it their task to create a research basis for these various professions to the farthest extent and as rapidly as possible and to develop that basis into quasi-disciplines with training programs, higher degrees, learned associations, journals, and textbooks. Scientists in the established disciplines had many misgivings about the danger of blurring the borderlines between disciplinary science and problem-oriented research which often lacked theoretical significance. In many cases this criticism was justified; the determination to conduct research relevant to the training functions of the university resulted, at times, in research which was irrelevant theoretically as well as practically.[10] What needs emphasis at this point, however, is that here, too, a function which was implicit in the state of science in Europe but could not be properly fitted into the existing conceptions and organization of scientific work so that it could develop through a variety of exceptions and improvizations became defined, organized, and standardized in the universities of the United States.

This transformation occurred in the same way that the emergence of the graduate school in the basic scientific and humanistic disciplines did. Visiting Americans in Germany were not very sensitive to the invidious distinctions that existed in that country toward academically unrecognized (or not fully recognized) fields. For them a *Privatdozent* or an *Extraordinarius* doing interesting research in an institute or university hospital often seemed (as he often was) a pioneer rather than someone specializing in a field that made him *nicht ordinierbar*, that is, unsuitable for promotion to a chair.

The reason for these differences probably lay in the fact that, unlike their German counterparts, the American academics interested in the creation of more scientific professional schools did not in the beginning possess a monopoly of professional education such as existed in Germany.[11] Rather, they had to contend with a powerful British-American tradition of thorough practical training. Not only was this tradition defended by the survivors of the prescientific age in university faculties, but also by the freedom of the students to choose between types of university. Students insisted on being trained thoroughly in practice; they did

---

[10] *Ibid.* See also Fleming, *op. cit.*, p. 110 (about the superiority of training at Johns Hopkins). For a criticism of misconceived efforts to develop certain fields into quasi-disciplines, see Flexner, *Universities*, pp. 152–177.

[11] How this monopoly led to an erosion of the function of training for professional practice is amply illustrated by Paulsen, *op. cit.*, pp. 225, 261, 262–264, 269, 274–275, and 711–714.

not want to start learning this aspect of their professions after leaving the university.

As a result, the reform of professional education under the impact of modern science did not lead in the United States to the abandonment of the earlier tradition of learning how to do things through practical experience. The conception of scientific research that liberally included problem-oriented research was fully compatible with this practical orientation. Both the professional school and the graduate school of arts and sciences were conceived as places where students were trained for a particular professional practice, and both endeavored to bring the student to a point where he would be capable of working on his own.

Organized Research in the Universities

The introduction of graduate training in the basic scientific and humanistic subjects, and the active support of problem-oriented research related to professional training lowered the barrier against organized research in American universities. As the function of universities was to train people to perform and apply research of the highest standards, the universities had to have up-to-date research laboratories to make it possible. These facilities were not only necessary to make it possible for the professors to pursue their own research, but also for the training of graduate students. Furthermore, since the universities abandoned their qualms about training and research for practical purposes, there were few limitations on the kind of research functions that universities could engage in. Finally, the existence of a departmental structure in teaching probably made it easier to assimilate the administrative arrangements for research into the university.

In agriculture, education, sociology, and eventually in nuclear research the universities pioneered research on a scale that far exceeded the needs of training students and was, from the very outset, an operation distinct from teaching.[12] By 1900 the research organizations developed in some of the schools of agriculture, medicine, and even in basic scientific departments became a challenge to European science and served as an incentive for the establishment of new research organizations, such as the Kaiser Wilhelm Gesellschaft in Germany, and the British Research Councils. This development, then, was another function which had started in the German universities where professors had their small research institutes. But their growth within the European universities was curtailed by the rigidity of the university structure. When transferred to the United States, further growth of research institutes took place, which was then

[12] For what is in part a very critical description of the growth of institutes not related to teaching, see Flexner, *Universities*, pp. 110–124; for the opposite view, see Morrill, *op. cit.*, pp. 24–37.

146

partially imitated in Europe. But this imitation did not lead to comparable growth, nor did it take place within the universities. It led only to specialized nonuniversity research institutes.[13]

The Growth of New Disciplines:
Statistics as a Case in Point

The differentiation of higher education into three sections—undergraduate college, graduate school, and professional school—and the provision for research which was at times only loosely tied to teaching, opened virtually unlimited possibilities for the establishment of new fields. The growth of disciplines such as the social sciences, comparative literature, musicology, and other fields was indirectly stimulated by their popularity as undergraduate subjects. Undergraduate interest led to a demand for teachers trained in these subjects, and hence to departments, and in some cases, even to Ph.D. programs in these subjects. There was little risk, therefore, in initiating a disciplinary or quasi-disciplinary organization in a field that had some promise of either intellectual or practical value. Because of the immense variety of interests to which the university catered, there was a demand for teachers of an equally wide range of subjects. This in turn created a demand for postgraduate training, which in turn had a corresponding impact on the diversity and creation of departments.

The departmental structure reduced the risk of enterprise even further. New specialities could easily be accommodated and nurtured within existing departments—which always had a considerable degree of heterogeneity—until they were strong enough to operate independently.

A good example of this was the development of statistics. Both as a field of mathematics, and as a tool which could be applied to a very great variety of problems, statistics had a venerable history in Europe, going back to the seventeenth century. In the nineteenth century an important professional movement was initiated and led by Quetelet for the improvement and propagation of statistics.[14] Yet, as an academic field, statistics had remained a very marginal affair, and it did not develop a scientifically based professional tradition. The basic work done by mathematicians was usually unknown to the practitioners, and

[13] In Germany the most important organizations belong to the Max Planck (previously Kaiser Wilhelm) Gesellschaft; in Britain, to the various research councils; and in France, to the Centre National de Recherche Scientifique (CNRS). There are governmental and private research institutions in the United States, too, but they do not perform different types of research from the universities, and their share in the total research activity is smaller than in most European countries. Statistics are usually not comparable since research institutes financed through the ministries of education are usually included in the higher education sector.

[14] Terry Clark, "Institutionalization of Innovation in Higher Education: Empirical Social Research in France, 1850–1914" (unpublished doctoral thesis, Faculty of Political Science, Columbia University), pp. 19–21.

there was little continuity and coherence either in the theoretical or in the practical work.[15]

The reason for this state of affairs was that those who were most creative in statistics were usually mathematicians or physicists who were uninterested in changing their disciplinary affiliation. Or else they were amateurs interested in solving practical problems rather than in initiating fundamental research.

In order to make statistics into an academic discipline there would have had to have been a group of persons within the universities interested in identifying themselves as statisticians. These could have come only from those interested in the uses of statistics to the extent that they were capable of communicating with, and learning from, mathematicians interested in probability. Potential sources for the emergence of such a group were the geneticists, economists, social scientists, and psychologists aware of the statistical nature of their problems. But only an occasional few among these took a serious interest in statistics, since most important contributions in these fields consisted of experimental and observational studies where statistical methods played a relatively limited role. The advocates of quantitative methods were often among the relatively less creative persons in their respective professions, and the whole approach still had to prove its utility. Even where the utility was obvious and the statistical techniques involved were simple, there was no unequivocal evidence that more intensive statistical work would be the best way to improve the field. In the German academic system, therefore, where one person had to represent an entire established field, it was unlikely that he would be chosen with much consideration of his competence in the marginal subject of statistics.[16]

To the extent that chairs in statistics were established in Europe, these were stillbirths. These chairs came into being as a result of nonacademic pressure on the universities and not as a reflection of the converging interests of a number of sciences in the statistical method. Usually the universities resisted such pressure, but they were willing to make compromises in cases that were deemed academically unimportant, where there was a legitimate state interest involved, and where the subject could be kept at a distance from more important academic concerns. Since the law faculties were training grounds for prospective civil servants, there was a long tradition of providing courses in political science and administration in the law faculties. These were narrow courses of study with little academic standing and little practical utility. The field of statistics was added to these studies. Within the law faculty, statistics had little or no relationship to either mathematics or the biological and other sciences, which had a potential interest in it. Those appointed to chairs in the subject were

---

[15] Terry Clark, "Discontinuities in Social Research: The Case of the *Cours Élémentaire de Statistique Administrative*," *Journal of the History of the Behavioral Sciences* (January 1967), III:3–16.

[16] The best-known case to illustrate this point was the attitude toward Mendel's work by one of the most outstanding botanists of his time and the ensuing fate of his discovery. See Bernard Barber, "Resistance by Scientists to Scientific Discovery," *Science* (September 1, 1961), pp. 596–602.

148

usually persons who had their basic training in law.[17] Thus, whatever statistical work went on in Europe in and outside the universities, the chairs in statistics had little share in it and could not serve as centers for the emergence of a discipline.

The development in the United States contrasts sharply with this. The existence of flexible and expanding departments, with many more or less independent posts, made it possible for all the increasing variety of academic users of statistics in biology, education, psychology, economics, sociology, and so on to develop its own specialists in the field.[18] In the beginning the large majority of workers were too poor mathematically and had too narrow a view of their fields to do important work. By the 1920s there arose a growing awareness of the shortcomings and a demand for a sounder mathematical basis. Certain centers emerged for serious statistical work such as that at Iowa State University, which was stimulated by the needs of the agricultural research station connected with the university.[19] Still, for advanced training in the theoretical aspects of mathematics, the resources in the United States were insufficient.

Therefore, young American statisticians went to Britain, which in the twenties and thirties was the center of statistical research.[20] Having benefited from the British training, important centers for statistics arose in the United States during the late thirties, especially around Hotelling at Columbia and Wilks at Princeton.[21] They were later joined by several young Europeans who had

[17] Terry Clark, "Discontinuities in Social Research." For the situation in Germany, see W. Lexis (ed.), *Die deutschen Universitäten: für die Universitätsausstellung in Chicago* (Berlin: A. Ascher, 1893), Vol. I, 1893, pp. 598–603.

[18] Paul J. Fitzpatrick, "The Early Teaching of Statistics in American Colleges and Universities," *The American Statistician* (December 1955), X:12–18; James W. Glover, "Requirements for Statisticians and Their Training," *Journal of the American Statistical Association* (1926), XXI:419–424, which includes detailed information on the teaching of statistics in departments of mathematics, economics, and social science; in schools of business, education, and public health; and in psychology and agriculture.

[19] The leading figure at Iowa was Henry L. Rietz, a Cornell-trained mathematician, who prior to his appointment as professor of mathematics at Iowa had been professor of mathematics at the University of Illinois and statistician of the University of Illinois College of Agriculture for more than ten years. His first publication was a 32-page appendix to a treatise on breeding (1907); see *Annals of Mathematical Statistics* (1944), XV:102–104; F. M. Weida, "Henry Lewis Rietz 1875–1943," *Journal of the American Statistical Association* (1944), XXXIX:249–251. For a detailed description of the history of this center, see J. C. Dodson, "The Statistical Program of Iowa State College," *The American Statistician* (June 1948), II:13–14.

[20] Hotelling went to Rothamsted in 1929, and those working in London included Samuel S. Wilks (1932–1933) and Samuel A. Stouffer. For the beginnings of the movement toward mathematical statistics in the 1920s, see A. T. Craig, "Our Silver Anniversary," *Annals of Mathematical Statistics* (1960), XXXI:835–837.

[21] Ninety-five fellows of the Institute of Mathematical Statistics in 1967 received their doctorates in the United States. The largest numbers received their degrees from Columbia and Princeton—17 from each—followed by North Carolina and the University of California at Berkeley, with 9 doctorates from each. But many of those who did not receive their degrees from these universities were influenced in one way or another by these centers, especially by that at Columbia. This information is based on an analysis of data from *Statisticians and Others in Allied Professions* (Washington, D.C., American Statistical Association, 1967), and from *American Men of Science* (Tempe, Arizona: J. Cattell Press, 1962).

obtained their mathematical training in Central and Eastern Europe and in Britain.[22] During World War II, an additional impetus to the development of statistics was generated by the creation and operation of the Statistical Research Group.

This wartime cooperation had probably reinforced the sense of practicing a common and distinct discipline. It did not create this consciousness, however, which may be dated, at the latest, from 1935, when the Institute of Mathematical Statistics was founded. (The consciousness probably existed even earlier than that.) [23] Demands for the establishment of separate university departments of statistics were voiced at the meetings of the American Statistical Association. The first establishment of a separate department occurred at the University of North Carolina in cooperation with the state university of the same state, where, as in Iowa, there was important agricultural research interest in the subject. The establishment of this department was rapidly followed by similar establishments at other universities, including the most prestigious ones. This led to the enlargement of the number of practitioners of the statistical discipline and, with it, to the development of more theoretical work in the field, which has helped in its definition as an academic discipline.[24]

In addition to the departmental structure, which made possible the extension of statistical work in an increasing number of scientific fields, a crucial role in this development was played by the involvement of the university in training and research in applied fields. In the early decades of this century, statistics was primarily considered as a tool of applied research. Although Europe never considered this kind of research to be suitable for the universities, American universities undertook to provide for this type of work too.

This explanation is supported by the single significant parallel to the United States in the development of statistics, namely, Britain. In fact, as far as contributions to statistical theory go, those of the British were much more important than those of the American statisticians. Britain also preceded the United States in the establishment of the first chair in statistics, which was created at University College, London in 1933.[25] The theoretical superiority of the British work does not need a great deal of explanation. That country had a far more developed scientific tradition at the time and a less abstract school of mathematics than the United States.[26] Hence it was easier in Britain for a few first-rate minds to acquire the necessary mathematical background for statistics than it was in the United States.

As to social conditions for the development of the field, the first similarity

[22] A. Wald (Columbia), J. Neyman (University of California, Berkeley), and several others were among the foreigners who came to the United States during the 1930s.

[23] Craig, *op. cit.*

[24] The department of statistics at the University of North Carolina was founded in 1946–1947.

[25] The incumbent of the chair was Egon Pearson.

[26] For this information I am indebted to Professor Leo Goodman of the University of Chicago.

the professionalization of research in the united states

to be explained between the two countries is that in Britain there was a possibility to link statistical work done in applied fields with each other (especially in agriculture and biostatistics) and with academic work in mathematics. Unlike the situation in the United States, this was not due to the enterprise of the universities in the combination of different kinds of skills and interests for the purpose of training and research in applied fields. There was, however, a functional equivalent for bringing together people with relevant interests in the semiformal and informal networks and circles comprising the academic elite and outstanding researchers and intellectuals outside the academic field. As a result the work of "Student" (W. S. Gosset), which was done mainly in industrial research, and subsequently of R. A. Fisher, much of which was done at the Agricultural Research Council at Rothamstead, as well as the interest in biostatistics originating from the eugenics movement became related to each other and to a variety of academic work.[27] Thus there emerged, as in the United States, a consciousness of common professional interests.

The institutionalization of the discipline within the British universities was much slower and much more halting than in the United States, in spite of the priority of University College, London, in the establishment of a chair in statistics. This slower institutionalization is manifested in the circumstance that the foundation of the chair at London had nothing like the effect of the establishment of a department of statistics at North Carolina. The example was followed by other universities only after a great delay and after such departments had been proliferating in the United States, and one suspects, more under the influence of North Carolina than of the example of University College.

At the same time the single chair at University College was not a dead-end road like the earlier chairs on the Continent. Although this chair, too, was the result of external nonacademic influence on the universities, it was an academically and intellectually respectable innovation and not a lip service paid to the service functions of the university. This department had a very great influence on the growth of the discipline and was not isolated from any relevant scientific development.[28]

The main reason for this was the already mentioned existence of an informal system of interdisciplinary contacts bridging academic and practical research. But an important role was played in these contacts by the existence of a variety of people interested in statistics within different university departments,

[27] E. S. Pearson, "Studies in the History of Probability and Statistics, XVII. Some Reflections on Continuity in the Development of Mathematical Statistics, 1885–1920," *Biometrika* (1957), 54:341–355.
[28] Those who worked at University College in the 1930s included Egon Pearson, R. A. Fisher, and Jerzy Neyman, and prior to that, Karl Pearson, Yule, and "Student" (W. S. Gosset). In addition, many of the important statisticians went there for study and research. Those who actually taught there include 5 of the 15 persons named as the most important contributors to the development of present-day statistical method in the *International Encyclopedia of Social Sciences*. See M. G. Kendall in "Statistics: History of Statistical Method," *IESS*, 15:224–232.

namely in economics, mathematics, psychology, and some other fields. The greater continuity and cohesiveness of these peripheral interests in Britain than in Europe was again the result of a structural similarity with the United States. In Britain there was also a departmental system (though at a much smaller scale and of a much more hierarchic character than in the United States). It was therefore possible to develop within the universities a tradition of statistical research even without the existence of chairs in the field. Readers and lecturers in mathematics, psychology, demography, biology, and other fields interested in the statistical method could develop a small, but continuous and high quality tradition in the field, even before the establishment of a chair.[29]

This example shows that both the United States and the British systems have been better capable of developing within or in cooperation with the universities a field of research that originated from practical interests than the continental European systems. Both of the former systems could accommodate and continuously develop statistics as a quasi-discipline over a long period of time and thus nurture the emergence of mathematical statistics as a discipline.

Beyond these similarities, however, one is struck by the differences. In the United States, the universities played the decisive role in all phases of the development. Aware of the existence of practical needs, universities virtually initiated the quasi-disciplinary stage of statistics in agriculture, biology, economics, and so on. Out of this development they produced the disciplinary stage. Universities also played a decisive role in the movement for the stricter professionalization of the practice of statistics.

In Britain the part of the universities (i.e. members of faculties as well as the university as an organized body) played a much more restricted and passive role in the development. The university system was varied and flexible enough to cooperate with worthy amateurs and applied researchers. But the initiative was largely left to individuals outside the universities and, although there was considerable continuity in statistical work, there was little diffusion of the organizational innovation within the system. Based on a poorer scientific tradition than the British, the American development had, nevertheless, some inevitability about it. After the first few years, it is difficult to see how all the development might have come to a standstill. In England, however, such a standstill might have been possible, as the growth of the subject until well in the thirties was dependent on the cooperation between very few men, many of whom had no university connections.

The External Conditions:
Decentralization and Competition

The diffusion of innovations and the eventual assumption of its present multiplicity of functions by the American university has not

[29] In addition to those mentioned in Footnote 28, Charles E. Spearman, the famous psychometrician had also taught at University College from 1907 to 1931. See G. Thomas, "Charles Spearman," *Royal Society Obituary Notices* (1949), 5:373–385.

occurred as a result of a preconceived plan. In the formative years of the system—between the 1850s and about 1920—there was a wide range of ideas about the functions proper and improper to a university, and the arguments debated were in many cases the same as those debated in Europe. But the effects of these ideas were very different because of the difference between the ecology of American academic institutions and their European counterparts.

In Europe the procedure for university innovation was to convey the ideas to the government, which then rendered a decision among the conflicting viewpoints based on a more or less public debate of the issue.[30] In the United States, however, there was no central authority, or even informal "establishment," to lay down policy for the whole country. Therefore there was no concerted opinion on a national scale or organized action to press the government to put certain schemes into action, or at least support them. Rather, the protagonists of an idea tried to realize their schemes in institutions where they worked.[31] There were of course state-supported universities, as in Europe. But they were not the only ones, and they were far from enjoying monopolistic advantages. The most prestigious and the wealthiest universities were private corporations. Thus the system was far more decentralized than in Germany. There, different states competed with each other. In the United States the state universities not only competed among themselves; they also had to compete with the private universities.

Decentralization was not, however, the only condition which made the American system more receptive to innovations. An equally important condition was the absence of important monopolies conferred on the system as a whole. Early in this century, lawyers, doctors, teachers, and civil servants—to the extent that the latter were trained at all, except "on the job"—were often trained outside the universities. The most important middle class career was business, which did not at that time require either formal or certified training. The universities had to prove that they were useful and worthy of support by initiating new courses of study and research, and by successfully "marketing" their services.

The Internal Conditions:
The Structure of the
American University

Because the universities had constantly to adapt to innovations to maintain their standing and to compete for personnel and resources

[30] The debate was public in England, and there the government left a great deal of discretion to independent bodies like the University Grants Committee, the Research Councils, and the universities. See George F. Kneller, *Higher Learning in Britain* (London: Cambridge University Press, 1955). In France, too, there was public debate but no independent bodies, as shown in Chapter Six. In Germany, as seen in the last chapter, the central government started to take interest in science only in the 1870s, and there was much less public debate about science policy than either in Britain or in France.

[31] See Veysey, *op. cit.*, pp. 10–18, 81–88, and 158–159.

the professionalization of research in the united states

to do so, it was impossible that they should be run either in the civil service manner according to fixed personnel establishments and regulations, or in the manner of wholly autonomous corporations of teachers, scholars, and scientists. Hence the imitation of the German model did not involve the adoption of the German system of university government. The changes that occurred in this respect paralleled the changes that occurred in the organization of business. Until the 1860s college presidents were the managers of their institutions, acting on behalf of the trustees who formed the corporation that enjoyed legal ownership of the physical assets of the college. The rise of the new universities was accomplished by a new type of president who combined the qualities of autocrat, statesman, and entrepreneur. He was still very much the dominant figure, but increasing size, increasing complexity of task, and increasing self-esteem on the part of academic staffs of increasing eminence required that he be capable of delegating authority and acknowledging claims to academic freedom. This group of presidents nurtured the growth of the universities of the present day. They laid the foundation of the present-day structure of government by much less powerful presidents responsible to a board and assisted by a number of full-time academic administrators, such as vice-presidents, deans, and so on. The president had to be an entrepreneur modifying his policies and university organization in an ever-changing situation, and trying to push his university ahead in its class through careful forward planning and rapid exploitation of new ideas.[32]

In order to operate effectively under these conditions, the subunits of the university had to be (*a*) flexible enough to carry out all the diverse functions of the university as well as to adjust to new ones; (*b*) autonomous, so as to be able to make changes in courses of study, teaching arrangements, and staff recruitment without undue delay; and (*c*) of sufficient size to perform training and research functions effectively in fields requiring many kinds of specialization.[33]

The most important of the units which emerged was, and is, the department in the basic arts and sciences and the larger professional schools (the smaller professional schools are themselves departments). It is the American substitute for a European chair plus institute. However, instead of having a single person fictitiously representing a broad area of research, this task is given to a group that may actually represent the subject in its entirety.

This development had also occurred in Britain. There, however, the structure of the department was steeply hierarchical, with, usually, one professor directing the work of several juniors. In the United States the departments from the beginning were much more equalitarian because they comprised several teachers of the same rank. In Britain the authority of the head of the department extended even to scientific matters (e.g., in making decisions as to the kind of research which should be done in the department), and remnants of such

[32] *Ibid.*, pp. 302–311.
[33] *Ibid.*, pp. 321–332 on the development of departments.

154

authority still persist. In the United States the departmental chairman came to deal primarily, and then only, with administrative matters. With regard to research his task came to be to obtain provision for it from the central authorities of the universities and from outside patrons rather than to direct it intellectually.

The size of the American department and the presence of a number of professors within it made possible the growth of the department, and within the department the formation of independent research units composed of one or several teachers and graduate students. The size of the American department also made possible the introduction of relatively independent subspecialities without raising the question of what was within the discipline and what was outside it, and an increasing tolerance for interdisciplinary interests on the part of at least a few members of the department without seriously affecting the work within the discipline.

Institutes in the United States—unlike those in Germany which were established to facilitate the work of a single professor—are seldom attached to particular departments and practically never to particular professorships. They are often interdisciplinary ventures.[34] Their purpose has been either "mission-oriented" research to bring to bear the contribution of several disciplines on the exploration of a single problem (e.g., human development, urban studies, etc.), or to share a single piece of equipment (e.g., an accelerator), among different groups of research workers. The departments were well established as the basic unit of the university by the beginning of the century. Institutes began after World War I.[35]

The Results of the System:
The Professionalization
of Research

These developments have rapidly transformed the role of the scientist. By the first decade of the century there emerged the conception of the professionally qualified research worker. A Ph.D. in the humanistic or scientific subjects assumed the same meaning as the M.D. in medicine. Those who possessed the title were considered qualified for research just as an M.D. was qualified to practice medicine.

The requirement of a Ph.D. made suitable candidates scarcer, and raised, thereby, the market value of those who possessed the degree. But its principal effect was to create a professional role that implied a certain ethos on the part of the scientist as well as his employer. The ethos demanded that those who received the Ph.D. must keep abreast of scientific developments, do research, and contribute to the advancement of science. The employer, by employing a

34 Committees have also been interdisciplinary organizations, but they have been mainly for teaching and training and not so much for research.

35 See Flexner, *Universities*, pp. 110–111. Committees apparently did not exist in the 1920s, otherwise they would probably have been mentioned by Flexner. A search in the catalogs of the University of Chicago showed that they began to appear at that university in the 1930s.

person with a Ph.D., accepted an implicit obligation to provide him with the facilities, the time, and the freedom for continuous further study and research which were appropriate to his status.

This development was a new departure from the particular status of college teachers in the United States in the nineteenth century. They were then employees of presidents or trustees, both of whom were accustomed to treating teachers in a very authoritarian fashion, as if they were no more than the assistants of the president, helping him to do the job for which he was responsible. It also entailed an important departure from European usages. The role and career of the research worker was not one of the central elements of German science organization—which was the only one that mattered as late as 1900. There research was not considered a profession. In spite of all the growth of research within and outside the universities, the official acknowledgement and provision for the scientist's role had not changed throughout the nineteenth century. Scientific achievements were considered as being sacred, as expressions of the deepest and most essential qualities of a specially gifted person, which had nothing to do with institutional provision. Research was, so the fiction had it, a voluntary, nonpaid activity. A certain number of posts, mainly professorial, had something like an official charisma (*Amtscharisma*). Those who had this status also enjoyed very great freedom, few and relatively circumscribed duties, great honor, quite a good income, and complete security of tenure. These positions were not stages of an occupational career, and the freedoms and privileges attached to them did not carry over to scientists who did not hold such elevated positions. The professor was not in principle paid for research, but he occupied a role with a stipend which made it possible for him to do research as he wished. The *Privatdozent* could also do research if he could arrange it, but no provision was made for him to do it; [36] he not only received no salary, but he had no officially provided funds for research. If he worked in the laboratory, he did so on the professor's sufferance.

According to this view, research which was directly paid for was not considered as research because it had none of the metaphysical pathos of the deepest expression of a creative spirit. It was simple and bureaucratic work, which could be (and often was) as narrowly and specifically prescribed as the employer (such as the professor heading the institute) wished it to be.[37] Academic freedom in this scheme was the freedom of a privileged estate. This might have fitted the state of science early in the nineteenth century, when scientists were few and when amateurs still played an important role in science. But at the end of the century, when scientific research ceased to be an amateur activity, it was a poor and invidious way of ensuring the growth of science. At that stage, only an arrangement which combined regular employment with individual autonomy and scientific responsibility of the research worker could provide a satisfactory solution.

[36] See Busch, Alexander, *Die Geschichte des Privatdozenten* (Stuttgart: F. Enke, 1959), pp. 109–117.
[37] *Ibid.*, pp. 70–71.

the professionalization of research in the united states

The new conception of the scientist's role as a professional one, and the flexible structure of the university with its openness towards innovation also introduced a great many changes in the hitherto prevailing relationship between academic organization and science. Although American professors spend as much time on academic administration as their European counterparts, most of it concerns departmental affairs directly related to teaching, research, and personal matters in the field of most immediate interest to them. There is a much more selective involvement in divisional or university-wide affairs. Academics are active in these affairs not simply in their capacity as equal members, without differentiation, of a self-governing corporation. Academics in America become involved in administrative work because they are inclined to do administrative work and allow themselves to gravitate toward becoming administrators. They may participate in these affairs as experts who advise the dean or the president, in whom large powers still reside. Their tasks are parallel to those of the "staff" in other large organizations. Finally, they act as watchdogs of the autonomy of the academic staff to prevent the administration from doing anything that interferes with this autonomy. Here they appear as representatives of a professional body within a polycentric, pluralistic system of the allocation of power.

Such institutions as the senate and faculty assembly do not have much importance in the United States. Presidents are appointed by the trustees, although staff representations and consultations play an important part in determining who is appointed, and the deans are part of the administration and not elected heads of the faculties. The American professor is not legally a member of the university corporation. He has been from the very outset a professional, employed by an organization to perform certain loosely defined services. His loyalty to the organization often becomes very pervasive and deep, but it is also often limited by economic and professional considerations. He has (particularly in the period since 1945) regarded it as right to insist that the university he serves provide optimal conditions for the exercise of his scientific capacities and that it provide him with the freedom and the backing to establish those conditions for himself from funds which he seeks to obtain outside the university.

What is called academic freedom in the United States is not the autonomous self-government of senior teachers who act as a corporate body in directing the affairs of the university as a whole. Instead, it is the scientist's guarantee of freedom from interference with the direction of his work and the expression of his views by an administration representing a lay board and from interference originating from outside the university and mediated through the lay board and the administrators of the institution.[38]

The constitutional history of the American university is the history of the devolution of authority in intellectual and academic matters from the board of trustees and the president to the department and its individual members. This movement, coupled with the vigor of strong presidents, is the source of the

[38] See R. Hofstadter, and Walter P. Metzger, *The Development of Academic Freedom in the United States* (New York: Columbia University Press, 1955), pp. 396–412, for the development of the specifically American concept of academic freedom.

unequalled adaptiveness and innovativeness of the American university and the social structure of scientific research in America.

Some Results of the System

The emergence of the scientific role in the American university is intimately connected with mobility of American scientists, which in turn is the most important element in the adaptiveness of American universities to new possibilities in research and training. There used to be (and still is) great mobility in the German system also. But German mobility was strictly circumscribed by the structure of the academic career and the hierarchy of universities. People went from one place to another either to obtain higher rank or to be at a more famous university (which usually implied better facilities and a more attractive intellectual setting).[39] In the United States, there is, in addition, a great deal of mobility motivated by an individual's assessment of what he wants intellectually at a particular stage of his career or what he desires in income. Scientists may go from a high position at a first-rate university to a less prestigious university in order to get an institute or department, or improved facilities and conditions of work. Retired members of the most famous universities do not consider it beneath their dignity to go to teach at a small college. And similar considerations influence academics to move out of the academic system altogether. In connection with this, scientists have become less identified with their universities than with their discipline, although usually they very much prefer to work in the atmosphere of a university.[40] There exists a *professional community* of scientists or scholars in each field, and one's standing in this community is a more important matter than in other countries.

One of the tangible manifestations of the importance of the professional community is the relatively greater importance of professional-scientific associations in the United States than in continental Europe. They play a more important role in publications, their conventions are more important affairs, and there is a closer relationship between the scientific and professional aspects of their activities than in Europe (the British situation is closer to the American).[41]

Only in the United States has there been a general and early recognition that there is no necessary contradiction between creative accomplishment in research and the organization of research. This absence of prejudice against or-

[39] Zloczower, *op. cit.*, pp. 29–38.
[40] William Kornhauser, *Scientists in Industry: Conflict and Accommodation* (Berkeley: University of California Press, 1962), p. 71 ff.; Simon Marcson, *The Scientist in Industry* (New York: Harper & Row, 1961), pp. 52–57; and William Kornhauser, "Strains and Accommodations in Industrial Research Organisations in the United States," *Minerva* (Autumn 1962), I:30–42.
[41] The statement concerning journals is based on counts of physics journals at the University of Chicago Library and information from experts in other fields. It is interesting to note that one of the main *ideas* of C. H. Becker's unsuccessful reform attempts in Germany was to strengthen the influence of professional-scientific associations in higher education and science policy (see Erich Wende, *C. H. Becker, Mensch und Politiker* (Stuttgart: Deutsche Verlagsanstalt, 1959), pp. 110–113).

158

ganized research and its effectiveness through standardization made it much easier to devise increasingly complex and sophisticated types of organized research. Thus departments, research institutes, and laboratories soon outgrew their European counterparts in complexity as well as in size. By the thirties and perhaps even before, the difference reached a stage where in some fields European scientists were no longer able to compete effectively with their American counterparts.[42]

Research in Industry and Government

The rise of the scientific entrepreneurs and administrators, the professionalization of research careers, and the rise of standardized procedures for staffing, equipping, and costing different types of research made scientific research into a transferable operation. Administrators would move from university administration to the administration of large industrial or governmental research laboratories and establish research units of the same kind that existed in the universities. And research workers could work in any of these settings without having to change their professional identities markedly or give up their expectations or standards.

Of course the practice of scientific research in organizations that have nonscientific goals presents the possibility of conflict. Instead of pursuing intellectually promising leads, the researcher may be required to engage in less interesting scientific problems. He may, furthermore, be limited in his freedom to communicate and cooperate with his colleagues who work elsewhere so as to safeguard industrial or military secrets.

The attitudes developed in universities could not provide a ready-made answer to these problems, but they created a basis for a pragmatic approach to them. First of all they helped to make up a culture partly shared by industry and government which defined what could legitimately be expected of scientists. In this way, the culture of university science helped to create a congenial environment in nonacademic institutions for university-trained scientists.

In consequence, industrial research was given considerable autonomy and a long time span to show its creativity. The industrial research worker was not considered just as any employee to be assigned at will to all kinds of troubleshooting tasks. In these favorable circumstances, a type of research worker arose who was continuously and fully engaged in product development. Perhaps this role appeared first outside the university in the laboratory of Thomas A. Edison, where it was performed by self-educated inventors. Gradually, the role was assumed by trained scientists and engineers and became more integrated with the complex of activities regarded as falling within the jurisdiction of professional scientists.[43]

[42] For the superiority of physics laboratories and other arrangements for physical research, see Weiner, *op. cit.* For medical research, see Flexner, *Medical Education*, pp. 221–226.
[43] With the exception of that performed in a few large industrial research laboratories, there is little research performed in European industry. For the differences in total investment in general, and development work in particular, see Table I of the Appendix.

A very large variety of modes of supporting training and research by government and industry without direct involvement on their part in activities for which they are unqualified also emerged from extension of the research activity beyond the limits of the university. The most common are research or training grants, contracts and donations. The advantages are that they are (*a*) given to persons and organizations of proven competence; (*b*) give the recipients sufficient freedom to devise their own plans, and at times, even change their original scheme as soon as they find out that it is not the most fruitful one; (*c*) encourage constant reevaluation, criticism, and comparison of programs and changes in policy without the necessity of abolishing or drastically changing whole organizations.

The existence of professional research workers and standardized procedures for the organization of research have been necessary preconditions for this proliferation and flexibility of research activities. The close relationship between universities on the one hand, and government, business, and agriculture and the community in general on the other, had been initiated and managed by administrators specializing in academic and scientific affairs (university presidents, officers of foundations, governmental research directors). The emergence of the specialist in university and scientific administration with traditions in initiative and a considerable body of "know-how" has been a *sine qua non* of the recent growth of science in the United States.

A Comparison of
Scientific Organization
in the United States
and Western Europe

In western Europe, the new functions of science which emerged after the middle of the nineteenth century were grafted onto the national systems of higher education that had emerged in the first half of the century. In the national system, the universities—and in France, some of the *grandes écoles* also—were the centers of pure science. From the last decade of the nineteenth century they were increasingly supported from the budgets of governmentally financed research organizations and laboratories established from time to time in an *ad hoc* manner. Research aiming at the solution of practical problems, or more generally taking place in fields where likelihood of practical application was great, occurred in segregated and specialized research institutions. These were usually financed by government and directly accountable to it, but in a few cases the institutions were financed by industry. Finally, development work was done in industry, but only in a few instances was it effective and systematic. To make up for this deficiency, therefore, the governments in Western Europe since World War I, and more especially since World War II, have stepped into this field too, either by establishing applied research

160

institutions of their own, or by encouraging trade associations with direct or indirect subsidy to establish and operate such institutions.[44]

In the United States, the trend has been from specialized institutions of higher education to universities performing an increasingly greater variety of functions. And there has been a parallel development from relatively small-scale specialized research institutions to large-scale, multipurpose ones. Such developments occurred both in industrial and in governmental research institutions. In no case was this development foreseen or planned in advance. It was the result of trial and error within a pluralistic and competitive system. Nonetheless, as far as research is concerned, the superiority of large multipurpose organizations seems to have been demonstrated, and with it the hypothesis that research as a cooperative enterprise where ideas and skills can be indefinitely shared, and where the sources of stimulation are probably quite variable is superior to small and segmented institutions that cannot compete successfully with large and varied ones. In a large university there will always be some innovating fields and some generational change to ensure stimulation. However, in a small, specialized, and segregated institution the atmosphere may easily become extremely homogeneous. European experience supports this view. The liveliest places scientifically have been the capital cities, such as London, Paris, and at one time Berlin, and Vienna, which by virtue of the spatial proximity of many relatively small institutions provided the atmosphere that only very large organizations could have provided otherwise.[45]

Large, multipurpose institutions are particularly important in applied or "mission-oriented" research. Such investigations, with goals which are not derived from the normal internal processes of scientific research, are very likely to be interdisciplinary. Not only does the mission require it, but the attitude of indifference of the administrators toward the dignities of academic disciplines is also likely to favor it. Small, specialized research institutes are likely to be more resistant to multipurpose projects. Where the director and the senior staff are of the same disciplinary background, they are unlikely to seek new problems other than those which arise within the framework of their own disciplinary tradition. In a larger, more heterogeneous organization, the director is less likely to be committed to a particular discipline. Administrators who are interested in results but not in particular disciplines can greatly facilitate the process of bringing in new types of personnel and taking on new problems. Such changes will create crises in a small, specialized research institution. Some persons may have to lose authority or even their jobs in the process. Decisions, therefore, will be delayed.

Since the frontiers between basic and applied work are continually shifting,

[44] OECD, *Reviews of National Science Policy: France* (Paris: OECD, 1966), pp. 41–43; *United Kingdom, Germany* (Paris: OECD, 1967), pp. 60–66.
[45] Joseph Ben-David, *Fundamental Research and the Universities* (Paris: OECD, 1968), pp. 67–75.

the establishment of specialized institutions in a field which is promising today may immobilize resources at a future date when other fields have become more interesting. Here, too, the multipurpose research institution is more effective than one with specialized concerns.

American academic and scientific institutions have thrived because they have learned from experience. They had to learn from experience, since their mere existence was no guarantee of their eminence. They had to compete for fame through accomplishment, and they had to compete for funds and for persons. They were helped in this competition by administrators who were not bound by the results and reputations of particular persons and whose concern for the whole institution made them more open to the lessons of experience.

To a large extent this innovating function was and still is absent in Europe. Truly self-governing university corporations have rarely been able to exercise much initiative because of their tendency to represent the vested interests of their members. In effect much of their efforts have always been directed at preventing change and innovation.

Thus scientific policy making usually devolved on the government. As a result, policy was made at a great distance from its execution; and since it was always made for the system as a whole, there was little opportunity to evaluate its success, except by comparisons with other countries. Paradoxically, therefore, the nationalization of the university and the scientific research system which was supposed to lead to more objective and better coordinated planning of higher education and research has, as a matter of fact, debilitated the capacity of the systems to learn from experience. That situation has arisen because the centralized systems had no constitutive feedback mechanisms such as are given by situations where universities and research institutes are free to make innovations and compete with each other. Also, there was no room in these systems for the development of executive and entrepreneurial roles that would specialize in academic and research affairs, and which would not be too remote from the university's day-to-day activities and yet would also not be too completely absorbed into them.

The Balance of the System

The most obvious results of the system have been the transformation of the relationship between higher education and research on the one hand and the economy on the other. This enterprising system of universities working within a pluralistic, educational, and economic system has created an unprecedentedly widespread demand for knowledge and research and has turned science into an important economic resource.

One decisive question that we have not confronted so far is whether the system has also encouraged scientific creativity for its own sake. After all, even the most effective diffusion and use of science are not necessarily scientifically creative. New knowledge is created by very few people who are interested in

162

it and capable of creating it. And it has been believed by many that making the practice of scientific research into a professional career might inhibit scientists from freely following the paths opened to them by curiosity and imagination.

As a matter of fact, however, the widespread uses of science have created a very wide foundation for pure research, the aim of which is to increase knowledge without consideration for its potential uses. How the practical uses support science for its own sake can be seen from a comparison of the statistics of research expenditure in different countries. The support of all kinds of research per capita of the population or as a percentage of the GNP is greater in the United States than in Europe. Expenditure on basic research is a smaller fraction of the total national expenditure on research than in Europe, but the absolute sum spent on basic research in the United States exceeds by a very great margin the amount spent on it in other Western countries, and the same is true of per capita expenditure (Table 8–1). This table shows that entrepreneurial applied science, which extended research and training to new and often relatively risky fields, did not ultimately diminish the share of basic research relative to the society's total resources, as has been feared in Europe, but rather is associated with an increase of this share.

Furthermore, the widespread cultivation of applied research has not led to a loss in the autonomy of science, as was originally feared. Even though the public outlook that prevailed in the United States when the changes under discussion were initiated was of the kind that did not hesitate to judge research by the criterion of short-term utility, it was not forced on the scientific community by a central power or a single source of support. Rather, the job of creating the new kinds of institutions was left to academic and research administrators and policy makers, such as university presidents, heads and advisers of foundations, private industrialists, and some heads of government departments. Some of these were genuine believers in the value of pure science, others might have been true utilitarians believing only in the value of science applied to something else. But all of them had to face the two highly practical tasks of either earning money by research or obtaining it for research and higher education. In both cases they could succeed only if the research they sponsored or promoted was at a very high level, and they had to recruit and retain good scientists for this purpose. If they failed, the costs in terms of money and eminence were very high. They could never rest on their laurels; if they did they were forced into a condition of decline by their industrial and academic competitors.

It was learned that the best way to utilize science for nonscientific purposes was not through subjecting research or teaching to nonscientific criteria, but to aid it in its own immanent course and then to see what uses could be made of the results for productive purposes, for education, and for the improvement of the quality of life. The link between science on the one hand and industry and government on the other was not established by the industrialists or the civil servants giving instructions to scientists. Rather there has been a constant

Table 8–1 *

Gross National Expenditure on Research and Development in the United States and Western Europe Related to National Resources and Analyzed by Sector of Performance and Type of Research

	Absolute amount Million US $	Per capita US $	% as of GNP	SECTOR OF PERFORMANCE (% OF TOTAL)				TYPE OF RESEARCH (% OF TOTAL)		
				Business enterprise	Government	Other nonprofit organization	Higher education	Basic research	Applied research	Development
United States 1963–64	21,075	110.5	3.4	67	18	3	12	12.4	22.1	65.5
France 1963	1,299	27.1	1.6	51	38	—	11	17.3	33.9	48.8
Germany 1964	1,436	24.6	1.4	66	3	11	20	—	—	—
Italy 1963	291	5.7	0.6	63	23	—	14	18.6	39.9	41.5
United Kingdom 1964–65	2,160	39.8	2.3	67	25	1	7	12.5	26.1	61.4
Austria 1963	23	3.2	0.3	64	9	1	26	22.6	31.9	45.5
Belgium 1963	137	14.7	1.0	69	10	1	20	20.9	41.2	37.9
Netherlands 1964	330	27.2	1.9	56	3	21	20	27.1	36.4	36.5
Norway 1963	42	11.5	0.7	52	21	2	25	22.2	34.6	43.2
Sweden 1964	257	33.5	1.5	67	15	—	18	—	—	—

* Compiled from Organization for Economic Cooperation and Development, *The Overall Level and Structure of Research and Development Efforts in OECD Member Countries* (Paris: OECD, 1967), p. 14, p. 57 and p. 59.

and subtle give and take between professional scientists who had a fair idea of what they wanted and could do, and the potential users of science in the professions, industry, and government. This mutually advantageous interchange was established and has been kept alive by academic and research entrepreneurs acting ss organizers and interpreters between the interlocutors.

The economy has benefited from science, but a large enough proportion of the benefits has been ploughed back into research to ensure systematically organized pure research in an increasing number of fields. What began to appear in Germany around the middle of the nineteenth century, namely, a group of

the professionalization of research in the united states

workers, usually the students of a great innovator, concertedly working on a coherent set of ideas until they had exploited all its potentialities, has become the normal state of affairs in the United States. Because of their secure economic base—which was never established in Europe—these activities are now pursued in the United States regularly and in a constantly widening range of fields. Scientific growth, to the extent that it can be measured by manpower figures, resources invested in science, or publications, has been accelerating, with the United States setting the pace and forcing it on other countries, which have found it increasingly difficult to stay in the race. This is the positive side of the balance of the system.

Threats to the System

There are also, however, negative aspects. One of these is the delicacy of the balance between the internal structures and traditions of scientific and scholarly creativity and the demands of the economic and political powers. This balance is more delicate in the United States than elsewhere because the pluralistic entrepreneurial structure and expansive system of science and higher education require a much greater involvement of the universities in the affairs of society. This is the price paid for the greater support of science and scholarship.

Until the 1940s this involvement had typically taken two forms. Universities and colleges were at times pressed or impelled to institute degree courses in occupations that had practically no actual or prospective scientific content and to accredit courses of study with little intellectual content. A similar but more legitimate external influence led to the great extension of professional training at the universities in fields that had had a genuine, but still only potential and undeveloped scientific or scholarly content. The early efforts of land grant colleges in agricultural and engineering education, and the establishment of schools of education, business, social welfare, and several other fields belong to this category.

Judging from the vantage point of the present, these attempts have not caused serious long-term damage to the system. They were considered by the group of creative and devoted academic scientists, scholars, and administrators either as evils to be contained, or as challenges that spurred them on to extend serious research and study to these new fields. As a result, some of the worst anomalies have either been eliminated or contained without seriously diluting the quality of the system as a whole, and the professional schools have managed to raise the intellectual content of their curricula and have constantly grappled with this problem.

Again, it was the professional university administrators who played an important part in neutralizing the noxious consequences of such "service functions." The pressure for the institution of these intellectually problematic courses was exerted on (and in some cases initiated by) the university adminis-

165

trations. The scientists and scholars in the faculties of arts and sciences usually had little incentive or opportunity to become involved in these practical matters. Usually they regarded such involvement as a threat to science and scholarship. In this situation, university presidents had to act as mediators between the demands of the external environment (which drew the university toward greater involvement in the service of the community), and the requirements of the academic community, which demanded the greatest possible freedom for concentrating on pure science and scholarship. If we disregard small colleges that were serving particular local, religious, or ethnic groups, the greatest pressures were probably exerted by unenlightened state governments. These had the power, through their control of financial support, to force state universities into performing various nonacademic services. In principle, boards of trustees of private universities had similar powers, but in practice, at the most important private universities, they tended to share the academic rather than the non-academic outlook, at least in their capacity of trustees. Other groups, such as professional and voluntary associations, could only try to influence the universities by offering them support in exchange for the establishment of professional schools and the performance of similar services.

As a result, the leading private universities that were the center of the system had to contend only with a relatively limited amount of pressure to compromise their standards.[46] They were sustained by the prestige of their scholars and scientists, and by the fame of their institutions (not always exclusively intellectual). Despite the equalitarian deference system in the United States, intellectual eminence, relative economic independence, and faithful and well-placed trustees succeeded in protecting the autonomy of intellectual activity.

Since World War II the conditions have changed. There has been a great acceleration of what Weinberg has called "the force-feeding" of scientific growth.[47] This process had started earlier, and it began probably with the setting up of graduate schools in different professions. But since World War II these developments have been due to an overwhelming extent to the rapid rise in the central governmental support of science. The share of the federal government in total research and development expenditure grew from less than one-quarter in 1940 to more than two-thirds of the total in 1965.[48] Thus something similar happened in the United States to what had happened in Germany in the 1870s. Following the rise of a new system of research in the wake of a victorious war, the government assumed increasing responsibility for research.

[46] "Centre" is used here to refer to an integral part of a system which serves as a model for the rest. See Edward Shils, "Centre and Periphery," in *The Logic of Personal Knowledge: Essays Presented to Michael Polanyi* (London: Routledge & Kegan Paul, 1961), pp. 116–130; and "Observations on the American University," *Universities Quarterly* (March 1963), XVII:182–193.

[47] Alvin M. Weinberg, *Reflections on Big Science* (Cambridge, Mass.: The MIT Press, 1967), p. 106.

[48] OECD, *Reviews of National Science Policy, United States* (Paris: OECD, 1968), pp. 30, 33, Tables 1 and 3.

the professionalization of research in the united states

The reaction of the U. S. university system to the expanding opportunities was very different from the one that occurred in Germany. The universities took full advantage of the opportunity, and the allocation of funds by grants and contracts has preserved the decentralized and competitive character of the system. As a result, the U.S. universities have not lost ground to other types of institutions, such as institutes of technology and other specialized types of higher education, but, rather, have increasingly assimilated the latter into their own structure. They also increased their share in governmental research expenditure [49] (which is another contrast to what happened in Germany as a result of the entry of central government in the support of research).

But there are signs of crisis that perhaps can be attributed to an inflationary situation, where the scientific system, stimulated by central spending, has attempted to perform things that it is intellectually incapable of doing. One of the manifestations of this is the growth of research to an extent where serious questions are raised about its usefulness either from the point of view of its contribution to knowledge, or to the economy or to any other special social purpose.[50]

This in itself might only be a limited problem of waste that could be corrected. But it seems that the inflationary situation gave rise to further problems which make the correction of the situation difficult. The most acute of these is a new type of student problem. One of the distinctive features of the U.S. system has always been the willingness of graduates, especially of those who possessed only a first degree, to enter all kinds of occupations. This prevented the emergence of a significant group of university graduates who, either because of the specificity of their training or the level and content of their social aspirations, were unwilling to enter any but a few prestigious and well-remunerated occupations. The resulting existence of a large number of "unemployable intellectuals" had much to do with the alienation and radicalization of intellectual politics in Europe in the first third of the present century. This phenomenon has been virtually absent in the United States.

But this situation may not continue. The sudden rise of graduate education in fields for which there is no specific demand, and where the criteria of competence are not quite unequivocal, might have created the beginning of a problem of an excess supply of highly educated people, or at any rate a feeling among significant groups of students and graduates that they are not an integral part of society. This may be part of the background (in addition to the Vietnam War and the urban problem) of the present alienation and radicalization among U.S. students.

Because of further changes in the external situation and the internal structure of the university, it is difficult to assess the results of this development. The situation of the university in U.S. society has changed partly because it

[49] *Ibid.*, and pp. 33, 191, Tables 3 and 36.
[50] Weinberg, *op. cit.*, pp. 156–160; and Harold Orlans (ed.), *Science Policy and the University* (Washington, D.C.: The Brookings Institution, 1968), pp. 123–164.

the professionalization of research in the united states

now contains practically all the young people between the ages of 18 and 25 who are sensitive to and potentially active in public affairs. Although they are dispersed on hundreds of campuses, means of communication and transportation have greatly reduced the effective distances among them. Thus the U.S. student body became a potentially very great political force, similar to what students used to be in Europe and Latin America. In those countries political and intellectual life has been excessively concentrated in the capital and perhaps one or two other large cities. As a result the students who congregated in those places from all over the country assumed an importance in political activism that was nearer to their proportionate weight among the potential activists in those few cities, than to their weight in the total population. Thus universities became one of the most suitable centers of political activism. For the reasons explained here there is a similar situation in the United States today, in spite of the greater decentralization of political and economic life.

Whether this will lead to the politicization and the intellectual decline of the university, depends on the ability of the university to restore a strengthened sense of scientific purpose among its members, and a renewed balance between its research and training functions and the needs of society. It is possible, however, that even the regenerative capacity of the universities has been affected by the inflationary situation. The main strength of the U.S. university as an organization used to be its effective leadership. The recent developments have considerably eroded the authority and the responsibility of the university presidents. Generous funding given directly to the professors relieved the presidents of their function as the promoters and defenders of research at the universities. This weakened the loyalty of the faculty toward the administration. They ceased to see in the president an ally in the realization of the most highly valued part of their work. Furthermore, the exemption of this most important function from the central direction of the university has probably impaired in a more general way the sense of common purpose, and the ability to view the university as a whole among staff, students and administrators alike.[51]

It is difficult to predict what the final outcome of this fluid situation will be. The system may regain strength, or it may become the prey of politicization. Its present crisis has spread to many other countries, but not to all of them, and again it is impossible to predict whether this spells the beginning of a worldwide crisis of the scientific culture that emerged in the 17th century, or only of a new shift of the center of the scientific activity. It is not intended to fathom the future here, but in the next chapter an attempt will be made to identify the variables and their interrelationships which have determined the emergence of the present situation.

[51] *Ibid.*, pp. 101–111, 323–330.

the professionalization of research in the united states

conclusion
nine

1. The Social Conditions of
Scientific Activity

The central question of this book was: "How did scientific activity grow and assume its present-day structure?" Having been concerned in the previous chapters with the main stages of this growth, our study shall now attempt to summarize some of the general conclusions underlying the whole development.

The differences between scientific activity at various times and places were explained by two types of conditions: one was the changing constellation of social values and interests among populations as a whole which channeled the motivation of people to support, believe, or engage in science to different degrees. The second set of conditions was the organization of scientific work which was more or less effective in marketing the products of research and encouraging initiative and efficiency in it. Although the first set of conditions relates to social systems in the broadest sense, the second set becomes relevant when scientific work in a country becomes a relatively autonomous subsystem of society—that is, when people make a living from working as scientists, choose science as a career (or at least as an important part of their careers), or when the society seeks the services of scientists or scientifically trained people who are regularly employed in different contexts and who participate as a group in the political and ideological processes of that society. There is still a third level of conditions dealing

169

with the structure of the individual research establishment or with different aspects of the life of the scientific community, such as the social structure of different fields, associations, and so forth. In this book this last level of organization was treated only to the extent that it was relevant to the understanding of the place of science in society and, therefore, will not be taken up in this concluding chapter.

The takeoff into continuous accelerating growth of science was explained in terms of the first set of conditions. Between the fifteenth and the seventeenth centuries there arose influential groups of economically and socially mobile people in different places in Europe who were in search of a cognitive structure consistent with their interests in a changing, pluralistic, and future-oriented society. Empirical natural science (whose conceptual development was quite independent of these social circumstances) provided such a cognitive structure of testable validity. Although it did not provide anything like a logically and empirically satisfactory model for the explanation of social life, its constant advance aroused sufficient confidence in the belief that scientific methods would one day also provide the key for understanding man and society.

This constellation led to the emergence and the recognition of the scientist's role (see Chap. four). He was a person studying nature rather than the ways of God and man, and using as his intellectual tools mathematics, measurements and experiment instead of relying on the interpretation of authoritative sources, speculation, or inspiration. He was a person viewing the state of knowledge in his time as something to be constantly improved on in the future rather than something to be brought up to the standards of a golden age in the past. This new scientific role was recognized and accepted as equal in dignity and superior in the scope of its applicability to that of the traditional philosopher, theologian, or literary man.

Once the role of the scientist was established, there was a possibility that science could become a relatively independent subsystem of society. Still, until the middle of the nineteenth century, the differences in the growth of science among different countries continued to be determined mainly by the constellation of social values and interests in general, rather than by the incipient organization of scientific work.

The first signs of an independent subsystem began to emerge in the eighteenth century (see Chap. five). Absolutistic monarchs were apt to support science for its technological and economic implications, and did not wish to apply the scientific procedure of judging things by their results to political, religious, or even economic affairs. Nor did they wish to extend the application of universalistic standards to social and cultural affairs in general. Natural scientists started to become a professional community who took advantage of the opportunities offered to them from whatever quarter they came and turned them to the benefit of scientific inquiry and their own personal interests.

Organization became an important determinant of scientific activity around 1840 (see Chap. seven). After that the jumps in scientific activity

170

occurred as a result of the discovery of new uses of science leading to changes in the definition of the scientist's role; and innovation in research plant and organization. In each case of organizational change there was a country that served as the center and model for the innovation and from which the new role and organizational type was diffused to other countries.

Thus the government-supported academy and the employment of scientists in various educational and consultative capacities emerged in France. The combination of teaching with research in the role of the professor and the research laboratory ("institute") emerged in Germany. The trained professional researcher—the Ph.D.—and the department combining research and training and the more complex type of research institute that employs several senior researchers who may have different disciplinary backgrounds emerged in the United States (and partly in Britain). At each of these turning points the center of scientific activity shifted to the country where the innovation occurred, and innovations in the use and organization of science were eventually diffused to other countries, raising the general level of activity everywhere. Thus the rate of scientific activity—which determined the rapidity of the exploitation of the immanent potentialities of science—has been determined by a series of social innovations in the use and organization of science that occurred in different countries and have been subsequently adopted because they were considered as the optimal available patterns by the worldwide scientific community. The choice of the scientific community has been manifested by the convergence of scientists for advanced studies in certain countries and the tendency to copy the institutions of those countries. This is not to say that the model has been actually copied everywhere. That, of course, depended on broader social conditions (see Chaps. six and seven).

2. The Mechanism of
Organizational Change and Diffusion

Having established the general evolutionary pattern of scientific growth, our study has to show what the mechanism of selection of a certain type of role and organization was. It appears that this mechanism was competition between strong units of research operating in a decentralized common market for researchers, students, and cultural products.

That decentralized systems have been more effective in the production and selection of new types of roles and organizations than centralized ones, needs little comment. Like the perspective of science, the organization of work most appropriate for research is also constantly changing. Everything else being equal, therefore, a more decentralized system is likely to produce a greater variety of ideas and experiments than a centralized one. Due to the numerous unpredictable ways that science can be enjoyed and used, a greater variety of experiments conducted by those competing with each other is also likely to produce more widespread demand and hence greater expenditure on science than deci-

171

cisions made centrally by a few wise men. Decentralization and competition also provide a built-in feedback mechanism for distinguishing between what works and what does not work satisfactorily. Centralized systems have to create artificial mechanisms of self-evaluation which have not been too successful.[1]

As to the diffusion of innovations in scientific roles and organizations, it has been a less effective process than the diffusion of discoveries in substantive science. Organizational innovations made in small countries like Scotland or Switzerland had little direct impact on others. The Scottish universities were probably the best in the world for part of the eighteenth century, but they were not imitated anywhere, except perhaps in the United States, which at that time was a somewhat backward intellectual dependency of Britain.

The reason why small countries have had relatively little influence on the organization of science around the world has been the absence of effective international competition between the units of scientific organization. This in turn is because the mobility of teachers, students, and resources across national boundaries is difficult, and the languages of small nations are not widely known. Therefore the international diffusion of organizational and role models occurs not as a result of competition between equal units, but through the imitation of the innovations made in large countries. These latter are therefore much more likely to become scientific centers than small countries, and once they become centers they acquire monopolistic positions in science. France in the early decades of the nineteenth century, and subsequently Germany and the United States had monopolistic positions in advanced training and the evaluation of discoveries. In the case of Germany and the United States, they also had monopolistic positions in publications (see the Appendix). As a result, scientists from all over the world have made these countries their spiritual home and center. They adopted the patterns of work prevailing at the center, because many of them obtained their advanced training there and because the usages of the center in training, evaluation, and hierarchy of authority became the standard practice of the worldwide scientific community. In a situation where the organization of science (i.e. manpower and resources) is national, the internationality of science that welds scientists from everywhere into a community

---

[1] The U.S.S.R. experiment in central planning and direction of research used to be considered as ideal by many scientists in the 1930s. See J. D. Bernal, *The Social Function of Science* (London: Routledge & Kegan Paul, 1939), pp. 221–237. The principal argument in favor of this type of scientific policy was that it prevented waste and was coordinated with the economy. The relatively modest attainments up to that time could be accounted for as a result of the general backwardness of Soviet society. Now after longer experience and the investment of greater resources, it appears that this centralized system has still not produced as good results as decentralized ones. See Peter L. Kapitza, "Problems of Soviet Scientific Policy," *Minerva* (Spring 1966), IV:391–397. The centralized system did not even produce better means of measuring scientific output (which is a necessary condition of central direction) than those developed in decentralized systems. See E. Zaleski, J. P. Kozlowski, H. Wienert, R. W. Davis, M. J. Berry, R. Amann, *Science Policy in the U.S.S.R.* (Paris: OECD, 1969), pp. 37–47, 263–282, 457–486; and R. W. Davies and R. Amann, "Science Policy in the U.S.S.R.," *Scientific American* (June 1969), 220:19–29. The same applies to the similarly centralized French system.

172

centered in one country creates monopolistic advantages. As has been shown in the case of Germany, these advantages may prevent the shifting of the center to a new place, even after the usefulness of the patterns of the existing center has been exhausted.

3. The Financing of Research

As was pointed out in the first chapter, the question of how much should be spent on science in a given country was rarely discussed prior to World War II, as the total sums involved were small and insignificant.[2] Since the end of that war, however, the share of research and development expenditure in the gross national product and in general manpower has grown so rapidly that the question of how to determine the limits of this growth had to be raised.[3]

During this century centrally financed specialized research institutes have been established in all countries in fields considered practically important, such as health, agriculture, geology, and so on. In addition, an increasing variety of industries also found it advantageous to have research laboratories with broader or narrower scopes. At the other extreme, there are a few academies and other institutions in most countries that engage in research for its own sake, although in some countries the amount of research performed in such establishments is negligible.

There are, however, great differences among the various countries in (a) the degree of centralization of the financing and direction of scientific training and research, and (b) in the degree to which the functions of teaching and research are combined and performed by the same people and in the same organizations or by different people in different organizations. Although centralization and decentralization have many aspects, it is probably not too arbitrary to arrange the main scientific countries today on a scale of centralization and combination of functions.

The two great spurts in scientific production following basic changes in the organization and uses of science since 1840 took place in Germany and in the United States, that is, in large countries which had highly decentralized scientific systems and where the combination of research with higher education was maximal. It may well be that the combination of functions was not inde-

---

[2] One of the few who did discuss it was Bernal, *op. cit.*

[3] See, for instance, OECD, *Reviews of National Science Policy, United States* (Paris: OECD, 1968), p. 30, Table 1, about the growth of science expenditure in the United States. The total outlay on research and development in 1929 was 0.2 percent of the gross national product in 1940 it was 0.3; in 1941, 0.7; from 1946 to 1952, about 1.0; in 1956, 2.0; and by 1964 it was 3.0 percent. About the absence of and search for suitable criteria for deciding the optimum level of support for research, see Harold Orlans (ed.), *Science Policy and the University* (Washington, D.C.: The Brookings Institution, 1968), pp. 123–188; Zaleski *et al.*, *op. cit.*, p. 45; and Alvin M. Weinberg, "Criteria for Scientific Choice" and "Criteria for Scientific Choice II: The Two Cultures," *Reflections on Big Science* (Cambridge, Mass.: M.I.T. Press, 1967), pp. 65–100.

Table 9–1

Main Scientific Countries by Centralization of Science Organization and Combination of Teaching and Research

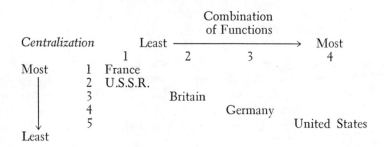

Centralization		Combination of Functions Least ———————→ Most			
		1	2	3	4
Most	1	France			
	2	U.S.S.R.			
	3		Britain		
	4			Germany	
	5				United States
Least					

pendent of decentralization. Higher education provides the most obvious and apparently the most numerous opportunities for the extension of the uses of science. Hence there is a likelihood that in decentralized systems where there is a great deal of initiative and enterprise in science, the delimitation of research from teaching will be constantly changing. Since more and more kinds of instruction will be linked with research and higher education, there will be a greater likelihood for the exploitation of the opportunities that higher education creates for research and vice versa.

On the other hand, the countries that had a centrally directed science policy tried to assess the needs for science and allocate funds for it accordingly. This assessment entailed an attempt to distinguish as much as possible between the different functions of science and created a tendency to organize and support each function separately.

Such policies, if executed by capable and wise people who have the support of the government, can be very successful in creating conditions for first-rate pure science by utilizing the experience of other countries. Experts in a field who know what happens elsewhere can make shrewd judgments about what is and what is not worthwhile to imitate and may have good ideas about improvements.

Such experts, however, will be in a much weaker position concerning the uses of science. Foreign experience can help in this respect too, but not to the same extent as in the case of science for its own sake where the objectives are always the same. There is no difference between what the physicist interested in atomic structure or the geneticist studying the evolution of a plant wants to know in Britain or in Japan. But if the question is one of what kinds of physical or genetical investigations may become economically useful in each of the countries or one of what kind and how much advanced physics and genetics should be used in the training of different kinds of experts in each of them, then the experience of the other country may be of relatively little relevance.

174

conclusion

A good example of this kind of research policy has been that of Britain during the last hundred years or so. Having possessed an established scientific elite with excellent political and social connections since the seventeenth century, it became alerted to the necessity of designing an official science policy earlier than any other country.[4] Much of this policy was practically motivated as shown by the debates surrounding its formulation as well as by the priority given to applied fields in the eventual establishment of research councils.[5] Nonetheless, the main success of these policies was in the basic fields. There are no indications of outstanding British success in applied science and development (although there is no basis either for the often-heard arguments about British failure in these areas). But in basic science, as shown by counts of publications, Nobel prizes, or any other index, British science has been quite exceptionally successful. It has maintained its place as the second country in science throughout all the changes in the contents and organization of science that have occurred since the eighteenth century, and its attainments throughout the whole period have probably surpassed those of every other country.

The explanation of these results is as follows. The scientific advisers of the British governments were probably as good as any and as far as science for its own sake was concerned, they were eminently capable of evaluating the lessons learned from foreign experience. But they were not competent in making economic choices necessary for practically oriented research, and even had they been competent, the examples known to them would have been irrelevant. Therefore all that they could contribute to applied research policy was the certainty that the work done in the research establishments was of high quality. In this they succeeded, so that these establishments have indeed made important contributions to science. But whether these contributions benefited the British or any other economy was a matter of chance.

Thus, contrary to its intentions, the British system became a model for a science policy designed to promote basic research. However, because the cost of basic research is also steeply rising, it is doubtful that this or any scientific policy which does not effectively promote the economic uses of science can be continued any longer.[6]

Scientific policy in other centralized systems has been similar to that in Britain. It has been based on the imitation of the scientifically most advanced country. But the quality of the advice obtained by governments has usually been worse, and the insistence of government bureaucrats on controlling scien-

[4] Pfetsch, Frank, *Beiträge zur Entwicklung der Wissenschaftspolitik* (Vorläufige Fassung), (Heidelberg: Institut für Systemforschung, 1969, unpublished mimeograph), pp. 23–26.

[5] See W. H. G. Armytage, *Civic Universities* (London: Ernest Benn, 1955), and D. S. L. Cardwell, *The Organization of Science in England* (London: Heinemann, 1957), about the debates on British scientific policy throughout the nineteenth century.

[6] Joseph Ben-David, *Fundamental Research and the Universities* (Paris: OECD, 1968), pp. 20, 55–58; C. Freeman and A. Young, *The Research and Development Effort in Western Europe, North America and the Soviet Union* (Paris: OECD, 1965), pp. 51–55 and p. 74, Table 6, for further discussion and information about the international differences in the economic benefits from research.

175

tists has usually been greater in other countries than in Britain. As a result many applied research institutes arose that contributed neither to science nor to the economy. All these policies have been based on a common fallacy, namely, that the uses of science are transferable from one country to the other to the same extent as its contents. However, uses depend on complex social mechanisms that are rarely understood and usually not taken into account in processes of diffusion and imitation.

The only case of centrally devised scientific policy not ostensibly based on imitation is that of the U.S.S.R. The main feature of the U.S.S.R. model has been the attempt of planning research as part of the central planning of the economy. But adequate criteria have not been found for determining the share of science in the economy. Furthermore, recent studies indicate that research and development played a relatively small role in the growth of the Soviet economy.[7] These facts confirm the impression that, as a matter of fact, the U.S.S.R., like other centralized countries, has pursued a policy of imitation to a much larger extent than may appear on the basis of its professed policies. Its great successes in applied science have been in the military field (and here too the pioneering was done elsewhere) where costs did not count, in basic research, and in science education. This indicates that, contrary to the intention of planning science in coordination with the economy, actual policy was largely based on the imitation of foreign models.

It appears, therefore, that there is no theoretically valid way to fix the share of scientific effort in the economy. The countries that tried to establish the level of support for science on the basis of comprehensive economic and educational policies have, in fact, taken as their frame of reference the situation in a few pioneering centers. In none of these latter was the level of support for science centrally established, but it emerged as a result of trial and error in decentralized competitive systems of decision making. Because science is a creative activity, and a means as well as an end, there can be no universally applicable standards to determine how much of it is adequate for a society. There can be only better or poorer mechanisms of regulating its level.

4. Problems in the Functioning of National Systems of Research

According to the present interpretation, the level of scientific activity in the centers of world science was set since the early 19th cen-

---

[7] See footnote 1. Also, about the factors of economic growth, see Raymond P. Powell, "Economic Growth in the U.S.S.R.," *Scientific American* (December 1968), 219:17–23. Powell does not treat research and development separately but shows that the increase in productivity did not exceed that of the United States during the last thirty-eight years. Since most of the increase must have been due to imported technology, the contribution of local R & D must have been considerably less than in the United States.

conclusion

tury by the mechanism of competition between the independent scientific units of these countries. This type of mechanism created a greater market for research and knowledge, and encouraged the emergence and the diffusion of more effective scientific organization than central direction of science. However, as the example of Britain shows, it has not necessarily ensured greater efficiency in the use of scientific resources. The remaining question is whether the level and type of scientific activity established by the competitive mechanism was optimal from the point of view of society.

To answer this question systematically would involve the formulation of criteria for the evaluation of the performance of national systems of research. Such a formulation will not be attempted here. All that will be done is to compare and interpret the difficulties in the functioning of the scientific systems that arose in Germany starting at the end of the nineteenth century and in the United States during the 1960s (see Chaps. Seven and Eight). The question will be asked as to whether or not the doubts that arose in both of these cases concerning the social justification and economic worthiness of the continued expansion of science and the dissatisfaction and moral disorientation of important parts of the scientific community were structurally related to the working of these competitive systems.

The details of the emergence of these crises indicate that there was probably no such relationship. Both in Germany and the United States one can distinguish between two periods. During the first period of initial growth the organizational innovations raised the levels of scientific activity above all previously attained levels. This came about practically without any interference by the central government. And the second period of accelerated growth was due to stimulation by central governmental support in Germany as well as in the United States. The problems occurred in both cases during the second period.

In both instances the central stimulation of scientific research came in the wake of victorious wars. It was motivated by a mixture of intrinsically scientific and extraneous military-political considerations. On the one hand, the successes accumulated in the previous period created numerous opportunities for accelerated scientific progress; on the other hand, there was a new recognition of the actual or potential military and political benefits of scientific research. The existence of new opportunities for science made it initially easy to decide what kinds of research should be supported and at what level. At the initiative of scientific lobbies the governments simply took over the programs that emerged at the previous stage of the development.

Once this stage of initial opportunities passed, the governments found themselves in a situation where they had to establish criteria for the support of science in the absence of either sufficient knowledge of the social consequences of science or foreign models to rely on as a frame of reference. In this state of uncertainty, the policy actually adopted was guided by a single purpose, namely the maintenance of the scientific supremacy of the leading country over all the

177

other countries. This modified completely the conditions under which the systems had operated in the preceding periods.[8]

The reaction of the two systems to this new situation was very different. In Germany, where the academic hierarchy was steep and authoritarian, and where the university was opposed to serving any utilitarian purposes, the policy adopted by the universities in the face of the new generosity of the government was "deflationary." The professors resisted any extension and diversification of the functions of the university and used the accelerated flow of research funds for the purpose of further enhancing the power and status gaps between themselves and other researchers. Thus in a situation of rapid increase of the resources and the prestige of the universities and science in general, the circulation of these resources was depressed below the level of opportunities inherent in the state of science and the demand for scientific services. As a result, the self-regenerative mechanism of the competitive system was virtually destroyed, as this became manifest from the inability of the German system to reform itself before the rise and since the fall of the National Socialist regime.

In the United States, where the university was not averse to regarding research and scientific training as an overhead on, or even means to practical ends, the reaction to massive government support was "inflationary." The system, which in this case was deliberately designed to exploit all opportunities, accepted increased government support as another such opportunity. Unwilling to forego any opportunity for expansion and diversification, the university undertook tasks beyond its capacity (or perhaps beyond any present-day capacity). This acceptance is said to have led to some misallocation of resources and may have contributed to the malaise felt at the American universities today.

It is possible to interpret the results of the deflationary situation in Germany between the 1880s up until the rise of the Nazis in 1933 and of the inflationary situation in the United States today in terms of Durkheim's theory of anomie. According to this, deflationary as well as inflationary crises give rise to social disorientation and despair, since the established criteria for judging the results of actions, as well as the customary relationship between means and ends of action lose their validity.[9] There are, indeed, some signs of such a parallelism. Despair, rebelliousness and resignation existed among professional scientists in Germany as a result of the difficulties and unpredictability of the scientific career; and there is a feeling of loss of purpose and dissatisfaction among scientists in the United States today as a result of too easy material success which is not always related to substantial contributions to science and society.

[8] The influence of Sputnik on the United States science effort does not require comment. The arguments used by Adolf Harnack in 1909 for the establishment of government research institutes in Germany related almost exclusively to the danger of German research being overtaken by others. See Adolf Harnack, "Zur kaiserlichen Botschaft vom 11 Oktober 1910: Begründung von Forschungsinstituten." *Aus Wissenschaft und Leben*, Vol. I, pp. 41–64.

[9] E. Durkheim, *Suicide* (New York: Free Press, 1951), pp. 241–245.

178

But one has to keep in mind the differences. The problems in Germany arose as the result of the impairment of the functioning of the competitive system by the monopolistic position of a small academic class. In the United States the present-day malaise started only after social and political problems, such as racial antagonism and the war in Vietnam, created tensions in the university. Thus while there is little doubt that the anomie in Germany was the result of imperfectionis in the working of the scientific system in the period following central interference (starting from the 1870s), in the United States the anomie appeared only when the system was also weakened by political and social conflict.

The difference is due to the better adaptation to research tasks and greater organizational strength of the U.S. system, and also the greater effectiveness of its decentralization. Still there is little doubt that the units of the system have been enfeebled as a result of force-feeding by the government. The leadership of the university, as well as that of the scientific community in general, lost much of their authority, and the far-reaching reliance on central government probably reduced the capability of the university to relate itself to the immediate social environment of research, such as the community, the parents, and the students. It is therefore an open question whether the system will regain its previous initiative as a result of the leveling off of governmental support or whether it will prefer to rely on political pressure to force the government into assuming complete responsibility for science. But whatever the outcome may be, this interpretation of the German and U.S. crises shows that they were not due to any inherent weaknesses of the decentralized competitive mechanism but to its impairment by the sudden rise of central governmental support to science guided by vague considerations of military superiority and national prestige.

5. Support of Science as a Means to an End and as an End in Itself

Apart from these conclusions, there are also other considerations which raise doubts about scientists' growing preference for support by central government. This preference is based on the assumption that because the benefits of science are shared by all, it cannot be expected that individuals or local groups will adequately support it. This argument is true as far as basic science is concerned. But in mission-oriented science the central government cannot be considered as the representative of the whole society, since it does not perform all the functions of society. If research relies mainly, or exclusively, on the central government, it will unduly favor central governmental functions, rather than all kinds of social needs that can be served by research. It seems, for instance, that the relatively large expenditures on military and agricultural research as compared to expenditures on research in housing and environmental hygiene—which is the case in most countries of the world—are at least in part due to the circumstance that scientific research, like defense and agriculture, is a function of the central government, while housing and environmental hygiene are affairs of local governments.

Furthermore, the argument about the social benefits from research does not take into account the fact that science itself has become an important economic enterprise. Scientists today are an interest group that competes for resources with other interested groups and thus may be involved in class conflict.

These new involvements of science with central government, military, and some industrial interests on the one hand, and the involvement of scientists in conflicts of class interests on the other, threaten the faith in science. Although, as has been shown in this book, much of the support for science was given for ulterior purposes, the belief in the ultimate moral usefulness of science was based on the belief that knowledge itself was a value. It is true that science has always been rather esoteric and inaccessible to the large majority of people. That statement seems to contradict the assertion about its importance as a cognitive value, but there has been a belief, supported by a great deal of actual experience, that the scientific method can be taught and widely applied and that it is a tool which is capable of improving the functioning of the human mind (if not actually its quality).

If science now comes to be regarded as serving some interests and not others, if it comes to be associated with military destruction, and if scientists are to be regarded as a privileged "clerical" group lavishly supported by government, then the faith in its values may be eroded by envy and doubt. Of course envy and doubt are irrelevant from the point of view of the cognitive validity of science. But they are relevant from the point of view of regarding the creation of scientifically valid knowledge as an end in itself. If science is perceived as partial to some social interests, and scientists are seen in an invidious light, then people may start doubting the moral value of seeking scientific truth for its own sake and applying it for the purpose of changing the world. This may spell the end of scientific culture.

In view of this, the virtually exclusive reliance of scientists on support by the central government may turn out to be a short-sighted policy. Although changing this reliance is easier said than done, scientific research will probably have to seek a much broader base of support and understanding than it presently has. Something that affects the fate of everyone as deeply as science can not afford to remain the affair of small circles of experts, civil servants, and politicians.

These last considerations lead us back to the relationship between social values and interests and science. Is it really true, as has been suggested in the last section, that an intrinsic interest and belief in science as a cognitive value is necessary for its survival? Or is it enough if society in general is interested in it for its technological applications?

At first sight it seems that the technological motive is sufficient. Science

180

survived and grew in several countries whose ideologies were apparently inconsistent with the values of science, such as fascist Italy, Nazi Germany, Stalinist Russia, and imperialist Japan.[10] Having taken up, or preserved science for military or technological reasons, these countries treated it in very much the same way other countries did. Furthermore, there have been scientists, even outstanding ones, who wholeheartedly supported autocratic regimes. The interpretation that this shows the autonomy of science from general social values can be supported by the argument that natural science is as value neutral as technology. Since it can be used for any variety of purposes, it is consistent with any system of values.

This argument, however, is erroneous. It is true that the correctness of scientific statements is independent of value judgments, but the decision to engage in or spend money on research is a choice between alternatives that reflects a scale of values. Governments that suppress many kinds of knowledge and still support science do this only for specific reasons. Some of these governments accept the values of science in principle and rationalize the transgression of those values as temporary measures made necessary by circumstances beyond their power to change. Where this is the case, scientists are in a relatively easy position, since they can see in their own work the "true" expression of the value system as opposed to the falsifications and imperfections of political and economic practice. This seems to have been an important element in the ethos of the scientists in the U.S.S.R. There the scientific situation today may in fact be parallel to the situation in France during different periods of the old and the Napoleonic regimes. The enthusiasm about natural science may well be rooted in the circumstance that this is the only field of cultural endeavor where the widespread belief in freedom, progress, and creativity can be spontaneously expressed.

Fascist and Nazi ideologies had no important elements of scientism. However, to the extent that these regimes continued to support science for military and technological reasons (and because of cultural and social inertia), scientism still survived as a value and as an important motivation among some of the scientists. They could regard science as a sacred haven of freedom where the brutal autocrat could not exercise his authority.

Science has survived in present-day autocracies because of its increasingly important military implications. This military importance imposes on the autocrats a great deal of restraint toward the scientific community, lest they fall militarily behind their potential enemies. Thus the relative freedom of physicists and theoretical statisticians under late Stalinism was probably not unrelated to

---

[10] For a statement about the independence of science from authoritarian politics, see Leopold Labedz, "How Free Is Soviet Science? Technology under Totalitarianism," in B. Barber and W. Hirsch (eds.), *The Sociology of Science* (New York: Free Press, 1962), pp. 129–141. For a theoretical statement of the opposite view which is followed here, see Robert K. Merton, "Science and Democratic Social Structure," in *Social Theory and Social Structure* (New York: The Free Press, 1957), pp. 550–561.

the recognition of the military importance of these fields. However, as shown by the policies of the Nazis and the Lysenko case, where a whole branch of science (genetics) was officially proscribed and its practitioners were persecuted in the U.S.S.R., if not for military competition, the prospects of science under such regimes could be very poor indeed.

6. Science and Social Values

But more can be learned about the relationship between scientific activity and social values from the cases of those societies where the dynamics of the situation can freely express themselves. It seems that in these countries there has been a cyclically changing relationship between science and scientistic social values. Periods when the issue of basic values is latent and those when research is accepted as valuable without any questioning, are followed by periods of opposition to the values of scientism and belittling of the importance of science. The romantic reaction to enlightenment, and many of the intellectual and ideological trends before both World Wars such as neo-romanticism, nationalistic populism, fascism, and so on, were all such instances. And just now there is such a period of transition from latency to questioning.

The causes of these cycles have not been systematically investigated, but their origin is indicated by the common underlying theme of impatience with the inability of science to dissolve basic human anxieties and to solve all social problems. It appears irrelevant that scientists have rarely promised to deal with all social ills. They are the ones who are taken to task, because even without any intention on their part, science has a decisive influence on all the cognitive structures whereby man orients himself in the universe, in nature, and in society. Its impact began with the overthrow of biblical and classical astronomy, which used to place the earth in the center of the universe, and it continued with the overthrow of the views on the creation of the earth and man. And science's impact is perhaps strongest today as a result of its increasing mastery over illness and the consequent elimination thereby of one of the sources of permanent anxiety and hope around which so many religious practices centered. Science is also responsible for the creation of powerful tools that harness energy and modify the whole natural environment, and this has replaced the all-pervading anxiety about illness with a new anxiety about the possibility of the total annihilation of mankind on one hand and hopes for the mastery of space and man's genetic endowment on the other.

This constant change of the cognitive map of man as a direct result of science had an immediate effect on social and moral philosophy. Starting with the seventeenth century, a continuing process of scientization of moral and philosophical thought began. In its most extreme manifestations, the scientistic trend attempted a complete break with religion, traditional wisdom, and philosophy. It went beyond the bound of cognition, and it strove toward utopian transformation of society on purportedly scientific bases (Chap. six).

182

The recognition that this transformation was impossible served as the background for recurrent moral crises in modern societies. The impairment of the traditional religious criteria of truth on the one hand, and the recognition that science could not provide an alternative basis for a total world view, especially concerning matters of morality, opened the door for never-ending moral-philosophical speculations and experimentations. The contents of this quest are cognitively determined. But the quest is activated by a social mechanism. Important scientific discoveries, which for one reason or another raise utopian expectations about a cognitive re-ordering of the world, lead to heightened intellectual optimism. People are eager to learn, know, and improve themselves and their world. This happened in eighteenth-century France as a result of the popularization of Newtonian physics. It also happened in the late 1940s and the 1950s as a result of the utopian belief in the power of science to solve all the problems of mankind. Strangely enough, this belief was probably more the result of the creation of the atomic bomb by scientists during World War II than anything else. This may explain why the optimism was much less widespread in Japan than elsewhere. In the West the discovery of the bomb became a myth of the redemption of the free world through free science. In Japan, it was a disaster from the outset.

This upsurge of interest is frustrated externally and internally. The external frustration results from the inability of science to do all that people expected it to do, and in particular from the limited moral and/or practical benefit that many of the students and intellectuals derive from their studies and searches, which are motivated by utopian expectations rather than specific objectives.[11]

The internal frustration derives from the moral-philosophical speculations that are bound to be taken up in any period of heightened intellectual interests partly for intrinsic philosophical reasons and partly in response to the widespread popular demand for intellectual products. People who heard about Newtonian mechanics or about relativity theory realized that there were certain important mysteries potentially impinging on their lives that were impenetrable to common sense but understandable through science. This shook their beliefs in their traditions and made them susceptible to world views which claimed to have succeeded in re-ordering man's cognitive structure in view of the most up-to-date scientific state. The supply of such theories has been plentiful, since the undisciplined use of the fallout of scientific concepts has provided many opportunities for the construction of quasi-scientific philosophies, such as materialism, positivism, social Darwinism and all the other "isms." However unscientific these theories were, their failure was sufficient to legitimate the rise of logically equally poor world views of an antiscientific bias.

At first sight this development appears as almost inevitable. Freedom of thought and expression are basic to science, so one cannot advocate its abolishment at the point when it presents a threat to science. Besides, the problems

[11] For a somewhat exaggerated account of this phenomenon, see Joseph Schumpeter, *Capitalism, Socialism and Democracy* (London: Allen Unwin, 1952), pp. 152–153.

raised by these speculations seem to be real. Science is incapable of creating a morality, while it bears, at least indirectly, the responsibility of having impaired the traditional religious basis of morality. Hence doubts arise about the value and moral justification of the scientific pursuit among members of the scientific community too, and some of them are led to join the searchers for new meanings.

In point of fact, however, the impression of the inevitability of the process is based on a confusion of the logical problem with the social mechanism which turns it into a moral crisis. The circumstance that our knowledge is limited need not lead to a crisis. No one reached more extreme conclusions about the limitations of human knowledge than Hume, but neither he nor his intellectual environment arrived at a moral crisis as a result. They reacted to the conclusions as scientists and practical people do—by asking themselves what can still be done within the newly discovered limits.[12]

The triggering of the mechanism leading to moral crisis is, therefore, not an automatic consequence of the cognitive problem but a result of social conditions. When there is opportunity and scope for empirical social thought and social action, the frustration of exaggerated expectations from science and scientism need not lead to a crisis. In this situation there is a possibility of intellectual reorientation from impossible to possible tasks. Realizing that science did not have an answer to everything, social thinkers were willing to admit the importance of tradition (including religious tradition) in the maintenance of moral order and as a source of kinds of experience that science could not furnish. This approach to the moral and religious problem was more consistent with the norms of empirical science than the quasi-scientific attempts to create moral systems by speculation. Although there cannot be either objection to or support for such speculation on scientific grounds, the professional norm of scientific conduct requires the choice of empirically testable problems in preference to apparently more important, but unempirical ones, and it altogether rejects problems which appear to be insoluble.

To sum up, the conditions under which the pursuit of science without recurrent moral crises has been possible are:

(a) Political conditions that allow social experimentation and pluralism and that contain some methods for comprehensive institutional change and review of change without recourse to violence.

(b) A permanent attempt to extend scientific thinking to human and social affairs in order to formulate the problems of rapid cognitive and social changes caused by science and to devise empirically investigable procedures for dealing with those problems.

(c) The application of the professional norms of the scientist to the social thinker, which imposes the discipline of not discarding existing tradition except on those specific points where there is a logically and empirically superior alternative:

[12] Shirley Letwin, *The Pursuit of Certainty* (Cambridge: Cambridge University Press, 1965), pp. 59–71.

184

In societies where the first of these conditions is absent, there will be little chance for the development of social thought such as specified under (*b*) and (*c*). As a result waves of scientific enthusiasm in these societies are followed by vogues of antiscientism, romantic irrationality, and even antinomianism threatening the very existence of science. But the existence of conditions that allow social change is not a sufficient basis for the development of vigorous and disciplined social thought. Social conditions do not generate intellectual ability and moral responsibility; they only provide the conditions for their exercise.

appendix

I. There is no satisfactory way of measuring scientific output, but there is almost complete consensus among historians of science about the existence of centers and the shifts which have been described here from Italy to England in the middle of the seventeenth century, from there to France during the second half of the nineteenth century, then to Germany about the middle of the nineteenth century, and subsequently to the United States since the late 1930s.[1]

The only point on which some of the historians differ from this scheme is the place of Britain about the middle of the nineteenth century. According to some historians there was a brief period of British supremacy in science at that time. This impression is based on the outstanding achievements of Darwin and Maxwell.[2] But both the discussions of the state of science in England and in France at that time, and the life histories of scientists who studied at that time show that by the 1850s Germany was regarded as the scientific center.

The impression of British supremacy could have been due to a particular circumstance at the time. When the French decline began and German supremacy had not quite been established, Britain—which had steadily held second place in

[1] John Theodore Merz, *A History of European Thought in the 19th Century* (New York: Dover Publications, Inc., 1965), Vol. I, pp. 298–305; A. R. Hall, "The Scientific Movement and Its Influence on Thought and Material Development," in *New Cambridge Modern History*, Vol. X, pp. 49–51; and Chaps. 6, 7 and 8.
[2] H. I. Pledge, *Science Since 1500* (London: Her Majesty's Stationery Office, 1947), pp. 149–151.

186

science since the end of the eighteenth century—became relatively conspicuous.[3] There are indications of this in the tables that follow (see Tables 1–8, pp. 188–194).

II. Quantitative indexes, some of which are reproduced here, measure published papers or discoveries or discoverers. Papers, manpower, and money can be reliably counted; discoveries are more difficult to identify. But in counts of papers they are usually allocated to the country where the journal is published. This causes some distortions. Data on manpower are not available for most countries even for the recent past, and there is also a question of the definition of the relevant population. Historical information about expenditures on science is largely missing, too, and there are also problems of definition which make comparison difficult.[4]

Tables 1–6 and Figures 1–2 nevertheless give an idea and a fairly consistent picture of the changes that occurred. Figures 3–6 are an attempt at showing differences in the structure of the transmission of research traditions.

III. As an indication of the growth of scientific expenditure and manpower, Table 7 is a time series of expenditure on research and development related to the gross national product in the United States. By the 1920s the country with the absolutely and, probably, also relatively (i.e., as a per cent of the G.N.P.) heaviest expenditure on science was the United States. The negligible amount spent there on R & D in 1930 (160 million dollars; 0.2 per cent of the G.N.P.) shows that the cost of research could not have been a very important determinant of scientific growth until then.

IV. As to the relationship of the growth of science to that of the economy, Britain and France have been close to each other for the last 200 years and have had about the same long-term rate of economic growth (1¼% per annum). But one of the periods when the British rate of economic growth was almost certainly higher than that of France was the first phase of the Industrial Revolution, from 1780 to 1830, which coincides precisely with the period of French supremacy in science. The French rate of economic growth surpassed that of Britain between 1830 and 1860, that is, during the decline of French science (which was accompanied by a rise in privately supported technological education!).

In Germany scientific and economic growth were largely parallel. Economic growth started in 1834 with the establishment of the Zollverein and reached its peak in the 1870s and in the 1880s. By and large this coincides with the growth of science (cf. Chap. seven). But there has been no direct relationship between the two kinds of growth. Industrial and economic development in Germany was modeled on the British example; it caught up with Britain only a short time before World War I and has never attained the position reached by the United States.[5] In science Germany had surpassed all other countries by 1850 as a result of a development which was not modeled on anything else.

V. Table 8 shows the expenditures of different countries on research and development, personnel engaged in R & D, number of Nobel Prize winners, and

[3] Similarly Britain played a conspicuous role in the 1930s during the transition of the center from Germany to the United States. See the case history of statistics in Chap. 7.

[4] For a systematic discussion of the problem, see C. Freeman, "Measurement of Output of Research and Development: A Review Paper," UNESCO, January 1969 (stencil).

[5] This description of economic growth is based on W. A. Cole and Phyllis Deane, "The Growth of National Income," in *The Cambridge Economic History of Europe*, Vol. IV, Part 1 (Cambridge: Cambridge University Press, 1966), pp. 10–25.

187

the percentage of their contribution to the world production of chemistry and physics papers. As can be seen it does not indicate a clear relationship between inputs and outputs.

Table 1

The Number of Original Contributions to Physiology in Various Countries, by Five-Year Periods

	Germany	France	England	USA	Other	Unknown
1800–1804	4	2	6	—	2	2
1805–1809	1	3	2	—	2	—
1810–1814	3	7	2	—	—	—
1815–1819	9	8	4	—	1	—
1820–1824	9	10	2	1	3	—
1825–1829	20	7	4	—	—	1
1830–1834	21	6	5	—	5	2
1835–1839	25	10	4	—	3	1
1840–1844	38	16	7	—	1	1
1845–1849	53	6	6	—	3	1
1850–1854	52	11	5	—	3	4
1855–1859	74	26	3	—	3	—
1860–1864	82	15	—	—	10	—
1865–1869	89	1	2	7	1	—
1870–1874	76	9	2	1	5	—
1875–1879	79	5	9	1	8	3
1880–1884	49	5	10	1	15	3
1885–1889	39	5	13	—	13	3
1890–1894	65	7	15	3	16	4
1895–1899	54	5	18	6	15	7
1900–1904	78	2	14	11	18	4
1905–1909	59	5	28	5	11	1
1910–1914	66	6	24	9	9	4
1915–1919	20	1	9	14	4	—
1920–1924	47	2	13	24	8	2

SOURCE: A. Zloczower, *Analysis of the Social Conditions of Scientific Productivity in 19th Century Germany* (M.A. Thesis) Jerusalem, the Hebrew University. Based on K. E. Rothschuh, *Entwicklungsgeschichte physiologischer Probleme in Tabellenform* (Muenchen and Berlin: Urban Schwarzenberg, 1952).

188

appendix

Table 2

Number of Discoveries in the Medical Sciences by Nations, 1800–1926

Year	USA	England	France	Germany	Other	Unknown	Total
1800–1809	2	8	9	5	2	1	27
1810–1819	3	14	19	6	2	3	47
1820–1829	1	12	26	12	5	1	57
1830–1839	4	20	18	25	3	1	71
1840–1849	6	14	13	28	7	—	68
1850–1859	7	12	11	32	4	3	69
1860–1869	5	5	10	33	7	2	62
1870–1879	5	7	7	37	6	1	63
1880–1889	18	12	19	74	19	5	147
1890–1899	26	13	18	44	24	11	136
1900–1909	28	18	13	61	20	8	148
1910–1919	40	13	8	20	11	7	99
1920–1926	27	3	3	7	2	2	44

SOURCE: J. Ben-David, "Scientific Productivity and Academic Organization," *American Sociological Review* (December 1960), XXV:830. Based on a list of medical discoveries in F. H. Garrison, *An Introduction to the History of Medicine*, 4th Ed. (Philadelphia and London: Saunders, 1929).

Table 3

Discoverers in the Medical Sciences at the Age of Entering Their Professions (Age 25) in Various Countries, 1800–1910

Year	USA	England	France	Germany	Other
1800	1	7	8	7	4
1805	1	8	5	8	2
1810	3	11	6	6	2
1815	2	12	12	7	3
1820	3	11	23	18	2
1825	2	17	15	18	6
1830	8	12	25	10	6
1835	11	13	26	29	7
1840	5	24	22	35	12
1845	5	14	13	33	5
1850	10	18	21	37	10
1855	15	16	20	49	27
1860	16	23	13	61	23
1865	25	15	36	71	26
1870	25	15	31	83	41
1875	40	31	23	84	46
1880	48	17	40	75	50
1885	52	16	34	97	52
1890	43	11	23	74	41
1895	47	9	27	78	29
1900	32	9	17	53	30
1905	28	4	4	34	25
1910	23	6	7	23	18

SOURCE: J. Ben-David, "Scientific Productivity and Academic Organization," *American Sociological Review* (December 1960), XXV:832. Based on Dorland's Medical Dictionary (20th Ed.).

appendix

Table 4

Per Cent Distribution of References in Psychology Texts by Language

Text	Total		English	German	French	Other
Ladd, *Elements of Physiological Psychology*, 1887	100.0	(420)	21.1	70.0	7.4	0.5
Ladd and Woodworth, 2nd edition, 1911	100.0	(581)	45.6	47.0	5.2	2.2
Woodworth, *Experimental Psychology*, 1938	100.0	(1,735)	70.9	24.5	3.1	1.5
Woodworth and Schlosberg, 2nd edition, 1954	100.0	(2,359)	86.1	10.9	2.5	0.5

SOURCE: J. Ben-David and R. Collins, "The Origins of Psychology," *American Sociological Review* (August 1966).

Table 5

Number of Physical Discoveries (5-Yearly Sums). Heat, Light, Electricity, and Magnetism Made in Britain, France and Germany by Country and Period of Discovery

Period	NUMBER OF DISCOVERIES		
	Britain	*France*	Germany
1771–75	11	—	3
76–80	17	5	11
81–85	7	10	1
86–90	9	7	0
91–95	9	3	7
1796–1800	17	7	14
1801–05	32	16	26
06–10	17	14	11
11–15	22	22	15
16–20	17	69	12
21–25	32	57	22
26–30	22	34	32
31–35	48	21	32
36–40	51	48	58
41–45	48	59	50
46–50	45	107	88
51–55	48	69	101
56–60	51	48	122
61–65	36	66	109
66–70	33	58	136
71–75	82	74	136
76–80	120	88	213
81–85	124	150	286
86–90	180	199	419
91–95	141	154	443
1896–1900	186	206	525

SOURCE: T. J. Rainoff, "Wave-like Fluctuations of Creative Productivity in the Development of West-European Physics in the Eighteenth and Nineteenth Centuries," *Isis* (1929), 12:311–313, Tables 4–6.

appendix

Table 6

The Nationality of Nobel Prize Winners in the Sciences 1901–1966

	1901–1930	*1931–1950*	*1951–1966*
United States	5	24	44
Belgium	1	1	—
Canada	1	—	—
France	14	2	4
Germany	26	12	7
Japan		1	1
Netherlands	7	1	1
Great Britain	16	13	18
U.S.S.R.	2	—	7
Other Countries	22	17	—

SOURCE: *Encyclopedia Britannica* (ed. 1967), Vol. 16, pp. 549–551.

Table 7

Expenditure on Research and Development Compared With GNP in Selected Years (1930–1965) (in Billions of Dollars)

A	GNP	Total R&D Expenditure B	Total R&D Expenditure as % of GNP C
1930	90.3	0.16 [1]	0.2
1935	72.2	—	
1940	99.6	0.34	0.3
1941	124.5	0.90 [2]	
1945	212.0	1.52	0.7
1950	284.7	2.8 [3]	1.0
1955	397.9	6.3	1.6
1960	503.7	13.7	2.7
1965	681.2	20.5	3.0

[1] "Expenditure on Fundamental and Applied Research," estimated in *Science the Endless Frontier*, by Vannevar Bush (Washington, D.C.: U.S. Government Printing Office, 1945).

[2] Source of this series: Department of Defense, Office of the Secretary: see *Statistical Abstract of the United States*, 1960, p. 538. These figures are regarded by F. Machlup as underestimating actual disbursements by about 20 or 30 percent. See *The Production and Distribution of Knowledge in the United States* (Princeton, N.J.: Princeton University Press, 1962), p. 156.

[3] SOURCE: OECD, *Reviews of National Science Policy, United States* (Paris: OECD, 1963), p. 30.

Table 8

Comparison of Investment in Science and Scientific Output in Selected Countries

	Year	R&D Expenditure (millions of dollars)	R&D Expenditure, Per Cent of Total Spent on Basis and Applied Research (not including Development)	Qualified R&D Personnel Year	Number	Nobel Prizes 1951–1966	Percentage of World Production of Papers in: Chemistry (1960)	Physics (1961)
Britain	1964/65	2,155	38.6	1964/65	59,415	18	16 (Commonwealth)	14
France	1963	1,299	51.2	1963	32,382	4	5	6
Germany	1964	1,436	—	1964	33,382	7	9	6
Japan	1963	892	—	1963	114,839	1	9	8
U.S.A.	1963/64	21,323	34.5	1963/64	474,900	44	28	30
U.S.S.R.	1962	41,300 millions of old rubles		1962	416,000—lowest estimate 487,000—highest estimate	7	20	16

SOURCE: OECD, op. cit., 1967, pp. 14—Table 2, 59—Table 3: Joseph Ben-David, op. cit., 1968, p. 26; Wallace R. Brode, "The Growth of Science and a National Science Program," American Scientist (Spring–March 1962), 50:18. D. J. de Solla Price, "The Distribution of Scientific Papers by Country and Subject—A Science Policy Analysis," Yale University.

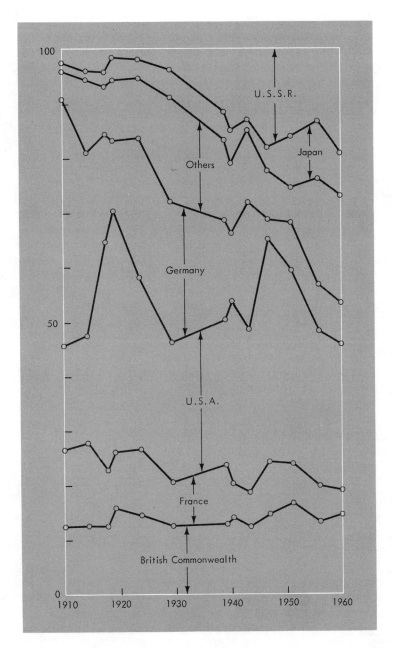

Figure 1. *Percentages of World Total Output of Papers in "Chemical Abstracts," by Countries, 1910–1960.* Source: *Derek J. de Solla Price,* Little Science, Big Science *(New York and London: Columbia University Press, 1963), p. 96.*

195

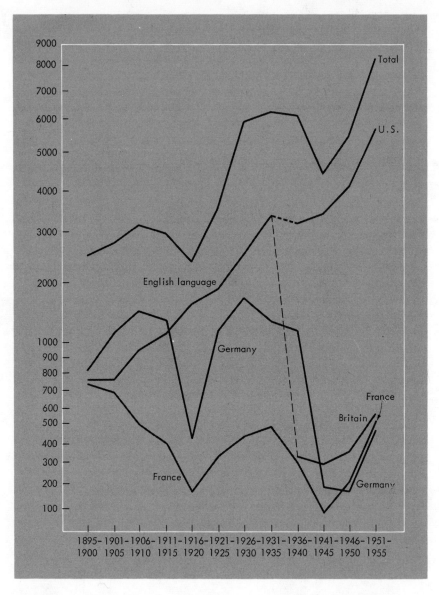

Figure 2. *Yearly Averages of Publications in Psychology, by Countries, 1896–1955.*

196

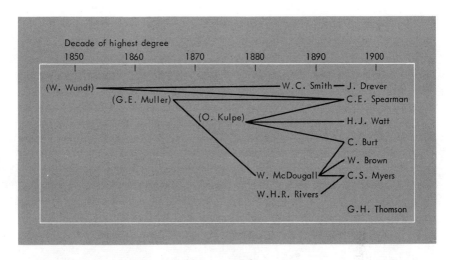

Figure 3. *Founders in Experimental Psychology and Their British Followers, 1850–1909.* Source: J. Ben-David and R. Collins, "The Origins of Psychology," American Sociological Review (August, 1966), 31: 4, 657.

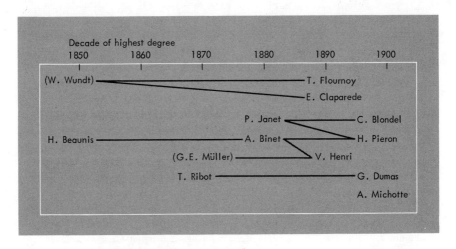

Figure 4. *Founders in Experimental Psychology and Their French Followers, 1850–1909.* Source: J. Ben-David and R. Collins, "The Origins of Psychology," American Sociological Review (August, 1966), 31: 4, 457.

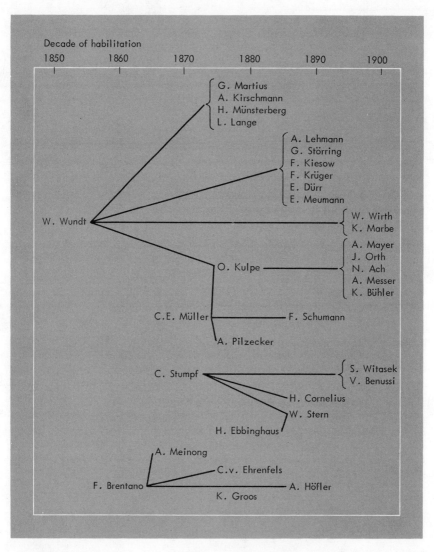

Figure 5. *Founders in Experimental Psychology and Their German Followers, 1850–1909.* Source: J. Ben-David and R. Collins, "The Origins of Psychology," American Sociological Review (*August, 1966*), 31: 4, 456.

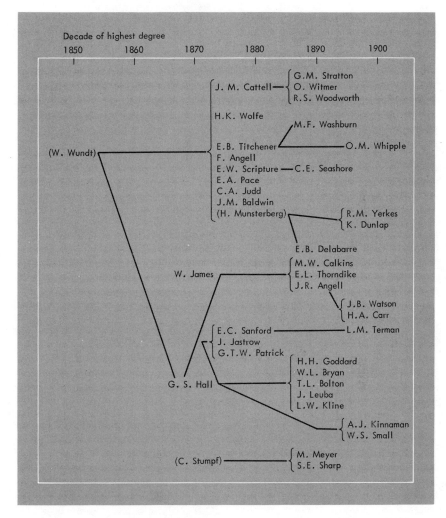

Figure 6. *Founders in Experimental Psychology and Their American Followers, 1850–1909.* Source: *J. Ben-David and R. Collins, "The Origins of Psychology," American Sociological Review (August, 1966), 3: 4, 458.*

index

Académie des Sciences, 66, 77, 78, 80, 82, 86; publishing rights of, 83
Academies, Italian, 59–66; development of, 59–60; "formal," 63; "informal," 63; science affected by, 63–66; status of science in, 61; universities and, 60–62
Accademia dei Lincei, 64
Affidati academy, 64
Aggregation, 121
Agricultural Research Council (Rothamstead, England), 151
Alberti, Leone Battista, 55, 56, 66, 70n
American Statistical Association, 150
Anatomy: 15th-century Italian artists and, 56, 57; medieval university studies in, 53, 54
Anaxagoras, 35
Appolonius, 39
Archimedes, 39, 58
Architects: early tradition of, 26; 15th-century Italian, 55, 58
Aristarchus, 39
Aristocracy, 67–68
Aristotle, 28, 37–40, 49
Artisans, 63, 67; 15th- and 16th-century Italian, 55, 57, 58
Artists, 15th-century Italian, 55–58
Astrology, 23; effects on astronomy by, 23, 24–25
Astronomy, 12, 13, 26, 182; Copernican, 64, 65, 71; medieval university studies in, 51, 52; practical tasks and scientific creativity in, 23, 24–25; Ptolemaian, 42–43
Athenaeum, 106
Atomic bomb, 183
Austria, 85
Auzout, 81

Babylon, 22, 24
Bacon, Francis, 58, 70; philosophy of, 72–74, 78n, 79, 81
Balbi, Bernardino, 56
Becker, Carl H., 137, 158n
Bentham, Jeremy, 90
Berlin, University of, 109, 115

Berthelot, Pierre, 106
Berthollet, Comte Claude Louis, 97
Bessarion, Johannes Cardinal, 59
Biggs, Henry, 66
Bildungsstaat, 137
Biology, 11, 53, 92
Black Death, 53
Bologna: medieval universities in, 47, 48, 52
Bonald, Vicomte Louis de, 9
Booth, Charles, 128
Borough, William, 67
Botany, 15th-century Italian artists and, 56, 57
Bourgeoisie, 8–9
Britain, *see* England
Brunelleschi, Filippo, 66; school of, 55, 56
Buddha, 34

Calendars, 23, 24, 25, 27
Cardan, Jerome, 56
Catholic Church, 63, 71, 94, 119; suppression of academies by, 64
Catholicism: empirical science and, 69
Centre National de la Recherche Scientifique, 106, 147n
Cesi, *Marchese*, 64
Chemistry, 11, 92; in German universities, 125–26, 131, 131n, 143; medieval university studies in, 53
China: early scientific traditions in, 22, 23; political-moral philosophy in, 38–39
Church and state: education controls by, 96–97; education problems due to separation of, 47–48
Cimento academy, 64
Colbert, Jean Baptiste, 81
Collège de France, 94, 101
Collèges, 94, 95
Colleges: land grant, 142
Columbia University, 149
Comenius, Johann Amos, 70, 80
Comte, Auguste, 9, 101, 128
Condorcet, Marquis de, 99
Confucian scholars, 28, 30n, 38
Copernicus, Nicholas, 13, 57; astronomy of, 64, 65, 71
Cusano, Niccolo, 57
Cuvier, Baron Georges, 97

Dalton, John, 89
Darwin, Charles, 11, 186
Davis, John, 66
Davy, Sir Humphry, 89

Dee, John, 66
Democritus, 38
Desargues, Gérard, 80
Descartes, René, 76, 80, 128
Diderot, Denis, 92, 93, 115
Digges, Thomas, 66
Donatello, Niccolo, 55
Dumont, Alfred, 106
Dürer, Albrecht, 56, 66
Durie, John, 72
Durkheim, Emile, 92, 128, 178
Duruy, Victor, 104n, 105

Ecole centrale des arts et manufactures, 101n, 104, 105
Ecole normale, 94, 101
Ecole polytechnique, 94, 95
Ecole pratique des hautes études, 103, 105
Ecole de santé (Ecole de médecine), 94
Economics, 127, 128, 131
Economy: capitalistic, 8; relationship of science and, 12–13, 14–16, 16n
Edison, Thomas A., laboratory of, 159
Egypt, early scientific traditions in, 22, 24
Eidgenoessische Polytechnik, 127
Electromagnetic theory, 7, 92
Elija, 35
Empedocles, 34, 35
Engineers: construction, early tradition of, 26; early social role of, 24; 15th-century Italian, 55, 57, 58
England, 15, 17–20, 75–87; change and reform in 18th-century, 90; development of statistics in, 149–52; 18th-century science in, 77–79, 83–85, 88; humanistic studies in, 95–96; institutionalization of science in, 75–77, 78, 80, 82, 83; intellectuals in 18th-century, 93; navigation in, 66–67; political and economic situation in 17th-century, 76–77; research policy in, 175; scientific discoveries in, 88n; social philosophers of 18th-century, 93; social philosophy and technology in, 79–80; support of Baconianism in Commonwealth, 72–74
Enlightenment, 93
Epicurus, 38
Erasmus, Desiderius, 58, 70n
Eratosthenes, 39
Euclid (Euclides), 39, 58
Eudemus, 37
Europe (*see also specific countries*): comparison of scientific organization in U.S. and, 160–62; Italian science and

Europe (*cont.*):
Northern, 66, 67–68; network of scientists and practical men in, 66–67; religious situation aiding science in, 64, 69–74; research expenditures in, 163, 164; science and class structure in, 58–59, 63, 65–66, 68–69
Evolutionary theory, 11
Extraordinarii, 144, 145
Ezekiel, 35

Faraday, Michael, 7, 11, 31, 89
Fichte, Johann, 115, 135n
Ficino, Marsilio, 59
Filarete, Antonio Averlino, 56
Fisher, R. A., 151
Fourier, Charles, 101
France, 15, 18–20, 88–107, 183; alienation of official education in, 93–94; conditions of reform in, 105–7; decline of science from 1800–30 in, 99–100; effects of institutional centralization in, 103–5; 18th-century science in, 77–78, 83–85, 88; flowering of science in, 99–100; humanistic studies in, 95–96; institutionalization of science in, 100–101; intellectuals in 18th-century, 91; reform of intellectual institutions in, 94–95, 113n; research in educational organizations in, 95–97, 124; science after Revolution in, 89; scientific discoveries in, 88n; scientism and science in 17th-century, 80–83, 89–94; scientists and practical men in, 67n; scientists' position after 1796 in, 98; scientists' position before 1789 in, 97–98; state control of education in, 96–97; structure of society in 17th-century, 81–82; universities in, 94
Francesca, Piero della, 56
Francesco di Giorgio, 56
Frederico, Prince (Urbino): court members of, 57
Frobisher, Martin, 66
Front Populaire, 106

Galileo Galilei, 57, 64, 65, 66, 70; private tutor of, 70n; Protestants and persecution of, 71–72
Gallicans, 81
Gassendi, Pierre, 76, 70
Genesis: creation story in book of, 28
Geometry, 51, 58
Germany, 15, 16, 18–20, 108–38, 181; academic freedom in, 118–19; academic self-government in, 118, 120–21;

203

index

206

207